Undergrad

P.R. Halmos

Undergraduate Texts in Mathematics

Anglin: Mathematics: A Concise History and Philosophy.
Readings in Mathematics.
Apostol: Introduction to Analytic Number Theory. Second edition.
Armstrong: Groups and Symmetry.
Armstrong: Basic Topology.
Bak/Newman: Complex Analysis.
Banchoff/Wermer: Linear Algebra Through Geometry. Second edition.
Berberian: A First Course in Real Analysis.
Brémaud: An Introduction to Probabilistic Modeling.
Bressoud: Factorization and Primality Testing.
Bressoud: Second Year Calculus.
Readings in Mathematics.
Brickman: Mathematical Introduction to Linear Programming and Game Theory.
Cederberg: A Course in Modern Geometries.
Childs: A Concrete Introduction to Higher Algebra.
Chung: Elementary Probability Theory with Stochastic Processes. Third edition.
Cox/Little/O'Shea: Ideals, Varieties, and Algorithms.
Curtis: Linear Algebra: An Introductory Approach. Fourth edition.
Devlin: The Joy of Sets: Fundamentals of Contemporary Set Theory. Second edition.
Dixmier: General Topology.
Driver: Why Math?
Ebbinghaus/Flum/Thomas: Mathematical Logic. Second edition.
Edgar: Measure, Topology, and Fractal Geometry.
Fischer: Intermediate Real Analysis.
Flanigan/Kazdan: Calculus Two: Linear and Nonlinear Functions. Second edition.
Fleming: Functions of Several Variables. Second edition.
Foulds: Optimization Techniques: An Introduction.
Foulds: Combinatorial Optimization for Undergraduates.
Franklin: Methods of Mathematical Economics.
Halmos: Finite-Dimensional Vector Spaces. Second edition.
Halmos: Naive Set Theory.
Hämmerlin/Hoffmann: Numerical Mathematics.
Readings in Mathematics.
Iooss/Joseph: Elementary Stability and Bifurcation Theory. Second edition.
James: Topological and Uniform Spaces.
Jänich: Linear Algebra.
Jänich: Topology.
Kemeny/Snell: Finite Markov Chains.
Klambauer: Aspects of Calculus.
Kinsey: Topology of Surfaces.
Lang: A First Course in Calculus. Fifth edition.
Lang: Calculus of Several Variables. Third edition.
Lang: Introduction to Linear Algebra. Second edition.
Lang: Linear Algebra. Third edition.

(continued after index)

Klaus Jänich

Linear Algebra

With 58 Illustrations

Springer

Klaus Jänich
NWF I-Mathematik
Universität Regensburg
D-93040 Regensburg
Germany

Jänich: Linear Algebra

First Indian Reprint 2004

ISBN 81-8128-187-X

This edition is published by Springer (India) Private Limited, Registered Office: 906-907, Akash Deep Building, Barakhamba Road, New Delhi – 110 001, India.

Printed in India by Rajkamal Electric Press, Delhi.

Preface

The original version of this book, handed out to my students in weekly installments, had a certain rugged charm. Now that it is dressed up as a Springer UTM volume, I feel very much like Alfred Dolittle at Eliza's wedding. I hope the reader will still sense the presence of a young lecturer, enthusiastically urging his audience to enjoy linear algebra.

The book is structured in various ways. For example, you will find a *test* in each chapter; you may consider the material up to the test as basic and the material following the test as supplemental. In principle, it should be possible to go from the test directly to the basic material of the next chapter.

Since I had a mixed audience of mathematics and physics students, I tried to give each group some special attention, which in the book results in certain sections being marked "for physicists" or "for mathematicians."

Another structural feature of the text is its division into laconic *main text*, put in boxes, and more talkative unboxed *side text*. If you follow just the main text, jumping from box to box, you will find that it makes coherent reading, a real "book within the book," presenting all that I want to teach. You might even learn linear algebra that way! But it is more likely that after reading the main text of a chapter, you will feel the need of a friend, someone who can explain things in more detail, give examples, and answer your questions. The side text is meant to be the voice of this friend.

Regensburg, Germany, May 1994 KLAUS JÄNICH

Contents

5. Matrix Calculus

6. Determinants

7. Systems of Linear Equations

8. Euclidean Vector Spaces

9. Eigenvalues

CHAPTER 1

Sets and Maps

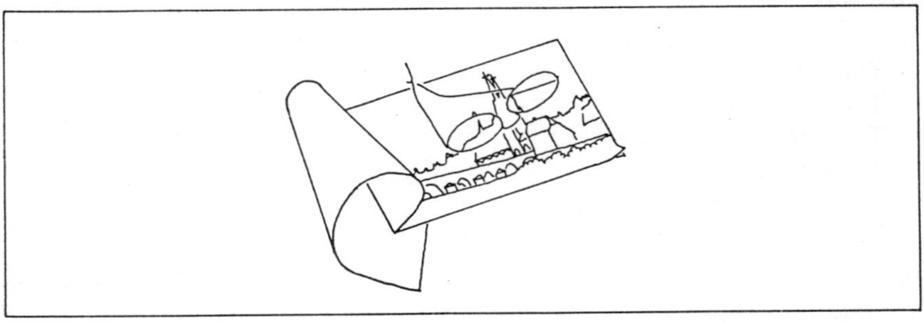

1.1 Sets

Throughout your entire mathematical study, and particularly in this work, you will be continuously involved with *sets* and *maps*. In an ordinary mathematical textbook these concepts occur literally thousands of times. The concepts themselves are quite easy to understand; things become more difficult only when we concern ourselves (from Chapter 2 on) with what in mathematics is actually done with sets and maps. First of all, then, let us consider sets. From Georg Cantor, the founder of set theory, comes the following formulation.

> "A ***set*** is a collection into a whole of definite, distinct objects of our intuition or of our thought, which are called the ***elements*** of the set."

A set consists of its elements. If one knows all the elements, then one knows the set. Thus the "collection into a whole" is not to be understood as doing something special with the elements before they can form a set. The elements form, are, and constitute the set — and no more. Consider the following examples.

$\mathbb{N}$ = the set of natural numbers = {1,2, ... },
$\mathbb{N}_0$ = the set of nonnegative integers = {0,1,2, ... },
$\mathbb{Z}$ = the set of integers,
$\mathbb{Q}$ = the set of rational numbers,
$\mathbb{R}$ = the set of real numbers.

The concept of a set consisting of no elements has turned out to be very useful. This is called the *empty set*, for which the notation is as follows.

$\varnothing$ = the empty set.

Next we introduce some signs and symbols, which one uses in connection with sets. Thus we have

The element symbol $\in$
The set brackets $\{\dots\}$
The subset sign $\subset$
The intersection sign $\cap$
The union sign $\cup$
The complementary set sign $\setminus$
The product set sign $\times$

Which among these signs is already known to you? What do they represent, when you simply make a conjecture from the names?

Let's look at the element symbol.

If M is a set and x is an element of M, then one writes $x \in M$. Correspondingly, $y \notin M$ means that y is not an element of M.

For example, $-2 \in \mathbb{Z}$, but $-2 \notin \mathbb{N}$.

For the set brackets, see the following box.

One can describe a set by writing its elements between two curly brackets. This writing out of elements can happen in one of three ways. If the set has only a few elements, then one can simply write them all down, separated by commas. For example, $\{1,2,3\}$ consists of the three numbers one, two, and three. Neither the order of the sequence nor whether some elements are repeated is of importance:

$$\{1,2,3\} = \{3,1,2\} = \{3,3,1,2\}.$$

The second possibility is to use periods to indicate elements that one does not write out. Thus $\{1,2,\dots,10\}$ is immediately understood to be $\{1,2,3,4,5,6,7,8,9,10\}$, and $\{1,2,\dots\}$ to be the set of all natural numbers.

However, one should only use this procedure when certain that each observer of the formula knows what the periods mean. For example, it would not be clear how to read $\{37, 50, \ldots\}$. The third, most frequently used, and always correct method is this: after the initial bracket, $\{$, one first writes a letter or symbol, which one has chosen to denote the elements of the set. One then makes a vertical line, on the other side of which one states in terms of this symbol, verbally or otherwise, what precisely *are* the elements of the set. Thus, instead of $\{1, 2, 3\}$, one can write: $\{x \mid x \text{ integral and } 1 \le x \le 3\}$. If the elements that one wishes to describe already belong to a specific set for which one already has a name, then one writes the property of belonging to the left of the vertical line: $\{1, 2, 3\} = \{x \in \mathbb{Z} \mid 1 \le x \le 3\}$. This reads: "The set of all x from $\mathbb{Z}$ with 1 less than or equal to x less than or equal to 3."

> To describe the third and most generally applicable way of using the set brackets, let E be a property that each x in a set X either has or does not have. Then $\{x \in X \mid x \text{ has property } E\}$ denotes the set of all elements of X which have the property E.

We use the subset sign as described below.

> If A and B are two sets, and if each element of A is also contained in B, then one says that A is a ***subset*** of B, and writes $A \subset B$.

Thus, in particular, each set is a subset of itself: $M \subset M$. Furthermore, the empty set is a subset of each set: $\varnothing \subset M$. For the sets introduced so far as examples, one has $\varnothing \subset \{1, 2, 3\} \subset \{1, 2, \ldots, 10\} \subset \mathbb{N} \subset \mathbb{N}_0 \subset \mathbb{Z} \subset \mathbb{Q} \subset \mathbb{R}$. In diagrams serving to illustrate the concepts introduced here, a set is often represented by a closed oval shape, labeled by a letter, as in Fig. 1. Then M is meant to be the set of points lying in the region "enclosed" by the oval. Sometimes, for the sake of greater clarity, we shall also shade the region where the points are elements of a set of interest to us. For example, in Fig. 2 we apply shading to indicate intersection, union, and difference (or complement) of two sets A and B. In case you are not yet acquainted with intersection, union, and complement, before reading further it would be a good exercise to try to understand the definitions of $\cap, \cup$, and $\setminus$ in terms of the pictures in Figs. 2a, b, and c.

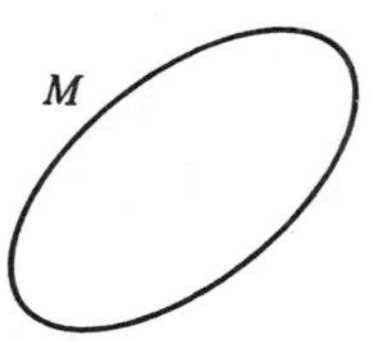

Fig. 1. A set M

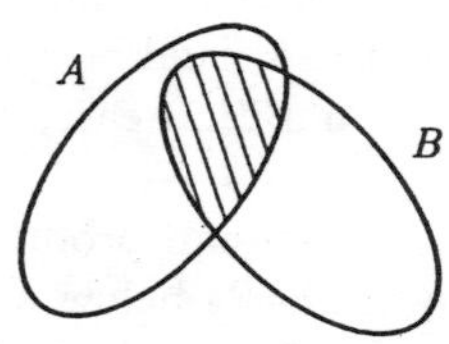

Fig. 2a. $A \cap B$

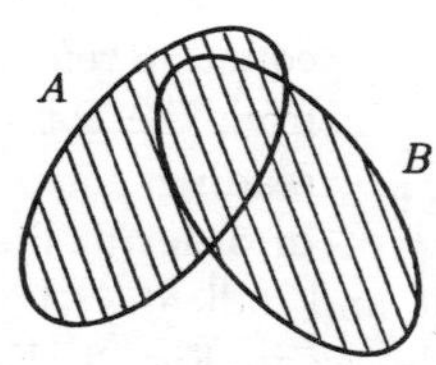

Fig. 2b. $A \cup B$

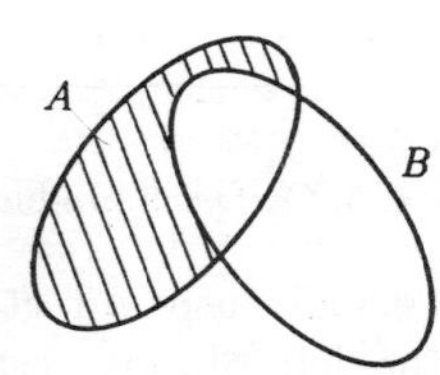

Fig. 2c. $A \setminus B$

Definition: If A and B are two sets, then the ***intersection*** $A \cap B$ (read as "A intersection B") consists of those elements that are contained in both A and B.

Definition: If A and B are two sets, then the ***union*** $A \cup B$ ("A union B") consists of those elements that are contained either in A or in B (or in both).

Definition: If A and B are two sets, then the ***complement*** $A \setminus B$ ("A minus B") consists of those elements that while contained in A are not contained in B.

When there are no elements "contained in both A and B," does it make sense to speak of the intersection $A \cap B$? Certainly! Then we have $A \cap B = \varnothing$, an example of the utility of the empty set. If $\varnothing$ were not admissible as a set, then in defining $A \cap B$ we would have to specify that there must exist some common element. Now think of what $A \setminus B = \varnothing$ means?

Before moving on to maps, we want to discuss Cartesian products of sets. To this end one must first define what is meant by an ordered *pair* of elements.

A ***pair*** consists in giving a first and a second element. If a denotes the first and b the second element, then the pair is denoted by (a, b).

The equality $(a, b) = (a', b')$ therefore denotes that $a = a'$ and $b = b'$. This is the essential difference between a pair and a two-element set: for the pair the sequential order is important, for the set it is not. Thus one always has that $\{a, b\} = \{b, a\}$, but $(a, b) = (b, a)$ only holds when $a = b$. A further distinction is that there exists no two-element set $\{a, a\}$ because $\{a, a\}$ has only one element, a. In contrast, (a, a) is a genuine pair.

Definition: The set $A \times B := \{(a, b) \mid a \in A,\ b \in B\}$ of pairs is called the ***Cartesian product*** of the sets A and B.

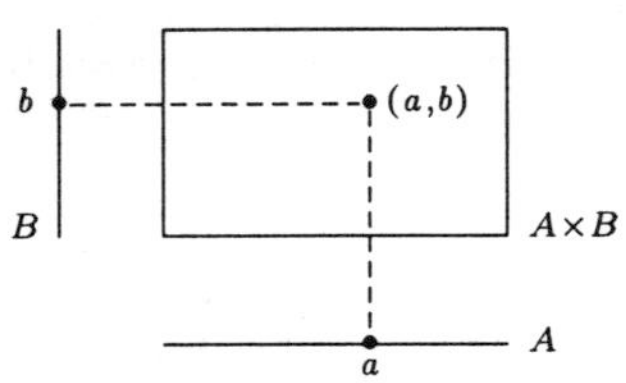

Fig. 3. Cartesian product $A \times B$

The symbol ":=" (analogously, "=: ") means that the expression on the side of the colon is first *defined* by the equation. Hence one does not have to search through one's memory to decide if one already knows it or to what the equation refers. Of course this should be clear from the context, but the notation eases its reading.

In order to illustrate the Cartesian product, one usually uses a rectangle and indicates A and B by intervals below and to the left of this rectangle, as in Fig. 3. For each $a \in A$ and $b \in B$, onethen "sees" the pair (a, b) as a point in $A \times B$. These pictures have only a symbolic significance: they illustrate the situation in a very simplified way, since in general A and B are not intervals. Nonetheless, as aids to thought and visualization, such diagrams should not to be discounted. One proceeds slightly differently when it is not a matter of consideringtwo sets A and B, but rather the special case $A = B = \mathbb{R}$. Here one "draws" $\mathbb{R}^2 := \mathbb{R}\times\mathbb{R}$ by sketching two mutually perpendicular copies of the real line. The horizontal line plays the role of $\mathbb{R} \times \{0\} \subset \mathbb{R} \times \mathbb{R}$, the vertical line that of $\{0\} \times \mathbb{R}$. An arbitrary element $(x, y) \in \mathbb{R}^2$ is then formed from $(x, 0)$ and $(0, y)$ as the diagram in Fig. 4 shows.

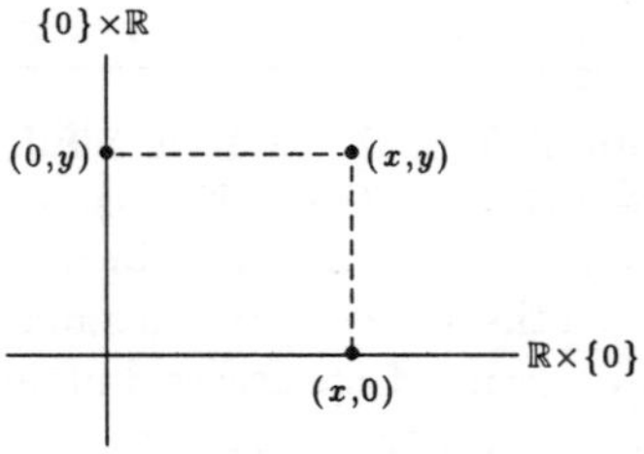

Fig. 4. Cartesian product ℝ×ℝ

Analogous to the definition of pairs, one also has ***triples*** (a, b, c) and ***n-tuples*** $(a_1, \dots, a_n)$. If $A_1, \dots, A_n$ are sets, the set

$$A_1 \times \cdots \times A_n := \{(a_1, \dots, a_n) \mid a_1 \in A_1, \ \dots, \ a_n \in A_n\}$$

is called the Cartesian product of the sets $A_1, \dots, A_n$. Particularly often in this book we shall have to do with the so-called $\mathbb{R}^n$; this is the Cartesian product of n factors $\mathbb{R}$:

$$\mathbb{R}^n := \mathbb{R} \times \cdots \times \mathbb{R} \, .$$

$\mathbb{R}^n$ is thus the set of all n-tuples of real numbers. Of course, between $\mathbb{R}^1$ and $\mathbb{R}$ there is only a formal distinction, if indeed one wants to make one. For the illustration of $\mathbb{R}^3$, as for $\mathbb{R}^2$, one uses the "axes" $\mathbb{R} \times \{0\} \times \{0\}$, $\{0\} \times \mathbb{R} \times \{0\}$, and $\{0\} \times \{0\} \times \mathbb{R}$, but we only half draw them; otherwise, the picture becomes difficult to read (see Fig. 5).

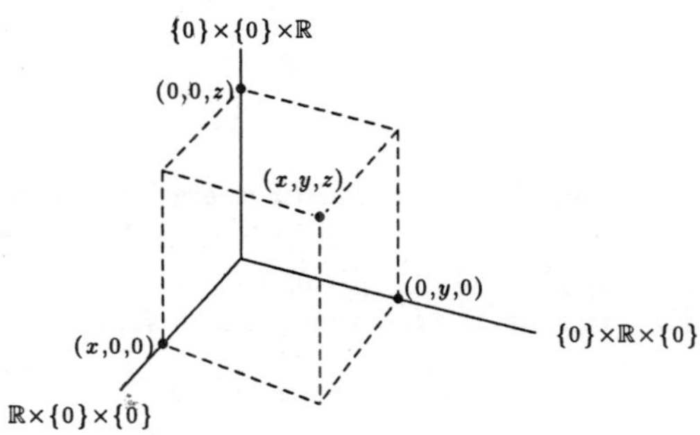

Fig. 5. Cartesian product ℝ×ℝ×ℝ

Such pictures are not meant to imply that $\mathbb{R}^3$ is "the space" (physical or geometric): $\mathbb{R}^3$ is and remains the set of all real number triples.

1.2 Maps

Definition: Let X and Y be sets. A ***map*** f from X to Y is a rule which to each $x \in X$ assigns precisely one element $f(x) \in Y$. Instead of "f is a map from X to Y," one writes $f : X \to Y$ as an abbreviation. Frequently it is practical also to describe the association of a single element x with its "image point" $f(x)$ by means of an arrow, but in this case, in order to avoid confusion, one uses another arrow, namely $x \mapsto f(x)$.

What does one write when defining a map? Here there is a choice of formulation. By way of example we use the map from $\mathbb{Z}$ to $\mathbb{N}_0$ that associates its square to each integer. Then one can either write

Let $f : \mathbb{Z} \to \mathbb{N}_0$ be the map given by
$f(x) := x^2$ for all $x \in \mathbb{Z}$,

or, somewhat shorter,

Let $f : \mathbb{Z} \to \mathbb{N}_0$ be the map given by $x \mapsto x^2$,

or, even shorter,

Consider $f : \mathbb{Z} \to \mathbb{N}_0 , \; x \mapsto x^2$.

Finally, it is sometimes unnecessary to give the map a label; then one simply writes

$$\mathbb{Z} \to \mathbb{N}_0 , \; x \mapsto x^2 ,$$

a very suggestive and practical notation.

One cannot avoid specifying which sets X and Y are involved (in our example, $\mathbb{Z}$ and $\mathbb{N}_0$), and it is also not permissible to call our map simply x^2. This is the value of our map at the point x, or as one also says, the *image* of x under the map, but not the mapping itself, for which we must choose some other notation.

Addition of real numbers is also a map, namely

$$\begin{aligned} \mathbb{R} \times \mathbb{R} &\longrightarrow \mathbb{R} \\ (x, y) &\longmapsto x + y. \end{aligned}$$

One can (and should) describe all arithmetic operations to oneself in this way.

A mapping does not need to be given by a formula; one can also describe the association in words. In order to distinguish between cases, one often uses a large bracket. For example, the function $f : \mathbb{R} \to \mathbb{R}$, defined by

$$x \longmapsto \begin{cases} 1, & \text{if } x \text{ is rational} \\ 0, & \text{if } x \text{ is irrational,} \end{cases}$$

is occasionally mentioned in analysis for one or another reason.

In a sequence of definitions, we shall now label some special maps as well as concepts and constructions referring to maps.

Definition: Let M be a set. Then one calls the map

$$\mathrm{Id}_M : M \longrightarrow M,\ x \mapsto x,$$

the ***identity*** on M. Sometimes one sloppily omits the subscript M and simply writes Id, if it is clear which M is involved.

Definition: Let A and B be sets. Then one calls the map

$$\begin{aligned} \pi_1 : A \times B &\longrightarrow A \\ (a, b) &\longmapsto a \end{aligned}$$

the ***projection on the first factor*** (see Fig. 6).

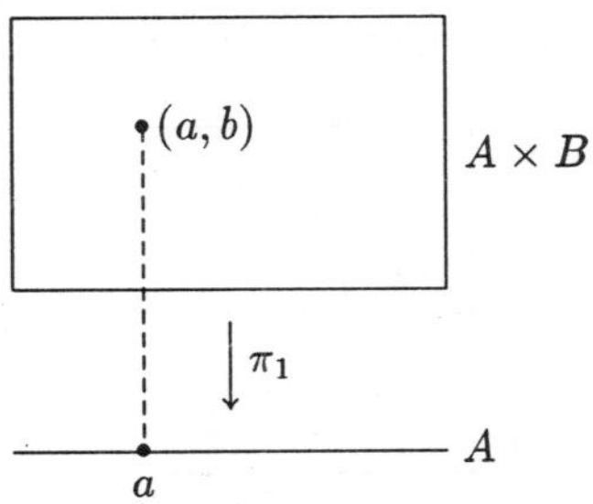

Fig. 6. Visualization of the projection on the first factor

Definition: Let X and Y be sets and $y_0 \in Y$. Then one calls the map

$$\begin{aligned} X &\longrightarrow Y \\ x &\longmapsto y_0 \end{aligned}$$

a ***constant*** map.

Definition: Let $f : X \to Y$ be a map and $A \subset X$, $B \subset Y$. Then one calls the set

$$f(A) := \{f(x) \mid x \in A\}$$

the ***image set*** of A or "the image of A," and the set

$$f^{-1}(B) := \{x \mid f(x) \in B\}$$

the ***preimage set*** of B or simply the preimage of B.

$f^{-1}(B)$ is read as "f minus 1 of B". It is important to observe that in no way have we defined $f^{-1}(B)$ using an "inverse map" f^{-1}. In this connection, the symbol f^{-1} alone, without an adjoining (B), has no meaning.

One can picture the concepts of image set and preimage set, shown in Figs. 7a and b, through the example of the projection onto the first factor of a Cartesian product given in Fig. 6.

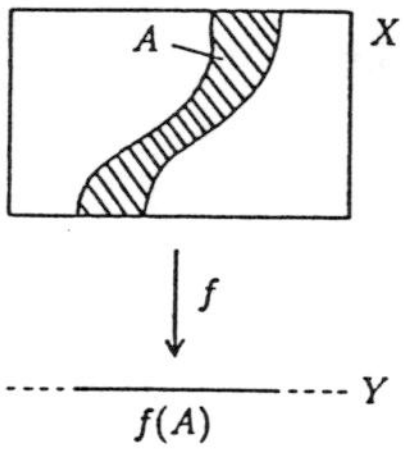

Fig. 7a. Image $f(A)$ of A

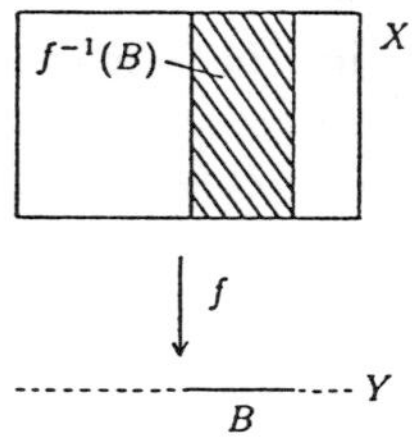

Fig. 7b. Preimage $f^{-1}(B)$ of B

The elements of $f(A)$ are precisely the $f(x)$ for $x \in A$. However, it can also happen that $f(z)$ belongs to $f(A)$ for some $z \notin A$, namely when by chance there exists $x \in A$ with $f(x) = f(z)$, as in Fig. 8.

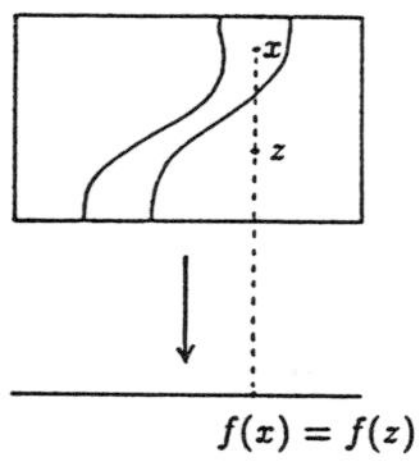

Fig. 8. It can happen that $f(z) \in f(A)$ for some $z \notin A$.

The elements of $f^{-1}(B)$ are precisely *the* elements of X that under the map f land in B. With maps it can also happen that *no* element lands in B. Well, then one has that $f^{-1}(B) = \varnothing$.

Definition: A map $f : X \to Y$ is called ***injective*** if no two elements of X are mapped onto the same element of Y. It is called ***surjective***, or a map ***onto*** Y, if each element $y \in Y$ is an $f(x)$. Finally, it is called ***bijective*** if it is both injective *and* surjective.

Let X, Y, Z be sets and f, g maps $X \xrightarrow{f} Y \xrightarrow{g} Z$. Then in an obvious way one can form a map from X to Z, which one writes as $g \circ f$, or in short form as gf:

$$\begin{array}{ccccc} X & \xrightarrow{f} & Y & \xrightarrow{g} & Z \\ x & \longmapsto & f(x) & \longmapsto & (gf)(x). \end{array}$$

The reason why one writes g first in gf (read "g following f"), even though one has first to apply f, is that the image of x under the composition of maps is $g(f(x))$. We formulate this as described in the following box.

Definition: If $f : X \to Y$ and $g : Y \to Z$ are maps, then the composition gf is defined by $X \to Z$, $x \mapsto g(f(x))$. If one has to deal with several maps between different sets, it is often clearer to arrange them in a diagram; for example, one can write the maps $f : X \to Y$, $g : Y \to Z$, $h : X \to Z$ in the form

$$\begin{array}{ccc} X & \xrightarrow{f} & Y \\ & {}_{h}\searrow & \downarrow g \\ & & Z \end{array}$$

If maps $f : X \to Y$, $g : Y \to B$, $h : X \to A$, and $i : A \to B$ are given, the corresponding diagram looks like this:

$$\begin{array}{ccc} X & \xrightarrow{f} & Y \\ h\downarrow & & \downarrow g \\ A & \xrightarrow[i]{} & B \end{array}$$

Of course, more sets and maps can occur in such a diagram; it is certainly clear enough what is meant by "diagram" without having to formalize this concept.

Definition: If in a diagram all maps between any two sets (including compositions and possibly multiple compositions) agree, then one calls the diagram ***commutative***.

The above diagram, for example, is commutative if and only if $gf = ih$.

If $f : X \to Y$ is a map and one would like to construct an "inverse map" from Y to X, which so to speak *reverses* f, then in general this fails for two reasons. First, the map f does not need to be surjective, therefore for some $y \in Y$ there possibly exists no $x \in X$ with $f(x) = y$, and hence one does not know which x to associate with y. Second, the map does not need to be injective, and therefore for some $y \in Y$ there may be *several* $x \in X$ with $f(x) = y$. But for a map $Y \to X$, only *one* x is allowed to be associated to each y. If, however, f is bijective, then there exists a natural inverse map that we can define as follows.

Definition: If $f : X \to Y$ is bijective, the ***inverse map*** to f is defined by

$$\begin{aligned} f^{-1} : Y &\longrightarrow X, \\ f(x) &\longmapsto x. \end{aligned}$$

One reads f^{-1} either as "f minus one" or as "f inverse."

Bijective maps will usually be denoted by the "isomorphism sign" $\cong$, thus

$$f : X \xrightarrow{\cong} Y .$$

Just as a precaution, let me add one further remark about the concept of an inverse map. Let $f : X \to Y$ be a map and $B \subset Y$ (see Fig. 9). You have just heard that only bijective maps have an inverse. However, experience shows that beginners are tempted to assume that *every* map f ought "somehow" *still* to have an inverse, and that $f^{-1}(B)$ has something to do with this inverse. I agree that the notation suggests this, but it should still be possible to distinguish between the bijective and nonbijective cases. When f is indeed bijective, then $f^{-1}(B)$ certainly has something to do with the inverse map, since you can describe it either as the f-preimage of B or as the f^{-1}-image of B. Clearly, one has (f assumed to be bijective):

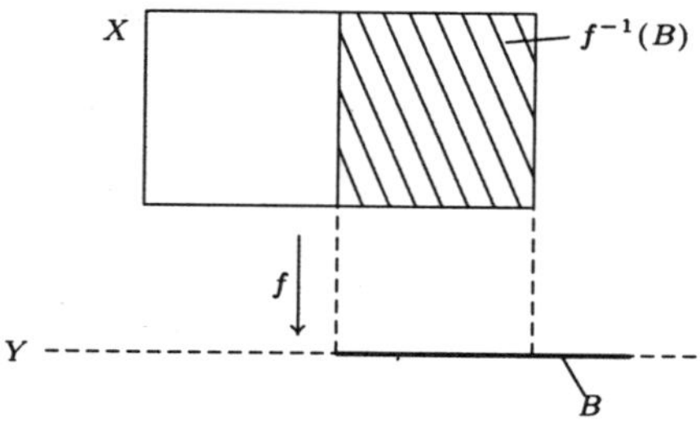

Fig. 9. Let $f{:}X \to Y$ be a map and $B \subset Y$

$$f^{-1}(B) = \{x \in X \mid f(x) \in B\} = \{f^{-1}(y) \mid y \in B\}.$$

One final definition: the restriction of a map to a subset of the domain of definition, shown in Fig. 10.

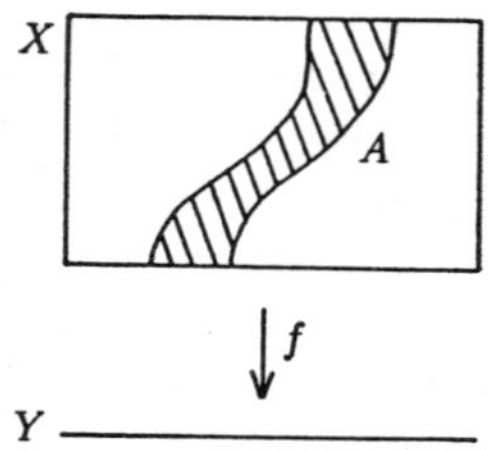

Fig. 10a. Map $f{:}X \to Y$

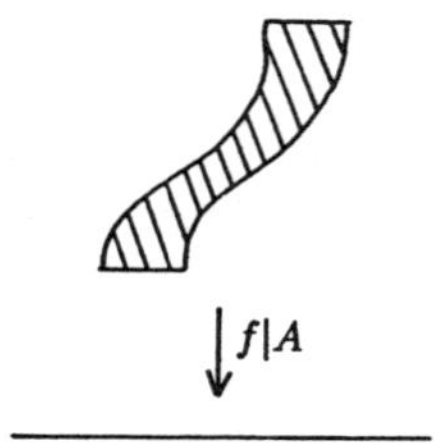

Fig. 10b. Restricted map $f|A{:}A \to Y$

Definition: Let $f : X \to Y$ be a map and $A \subset X$. Then the map

$$\begin{aligned} f|A : A &\longrightarrow Y \\ a &\longmapsto f(a) \end{aligned}$$

is called the ***restriction*** of f to A. One reads $f|A$ as "f restricted to A."

1.3 Test

(1) If for each $a \in A$ we have $a \in B$, one writes

☐ $A \subset B$ ☐ $A = B$ ☐ $A \cup B$

(2) For each set M, which of the following sets is empty?

☐ $M \cup M$ ☐ $M \cap M$ ☐ $M \setminus M$

(3) As usual, represent $A \times B$ by a rectangle. How would one picture $\{a\} \times B$?

☐
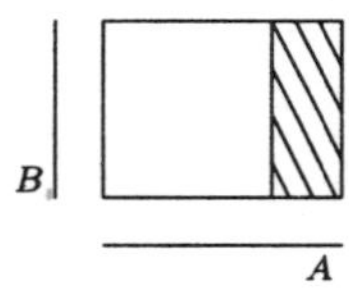

Fig. 11a.

☐
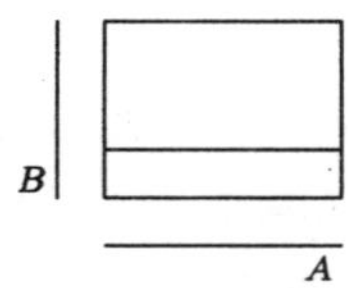

Fig. 11b.

☐
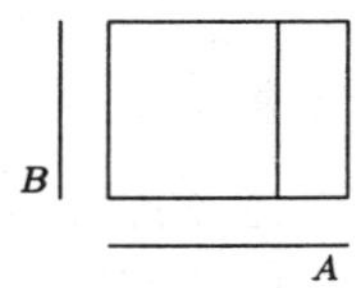

Fig. 11c.

(4) Which of the following statements is false? The map

$$\begin{aligned} \mathrm{Id}_M : M &\longrightarrow M \\ x &\longmapsto x \end{aligned}$$

is always

☐ surjective ☐ bijective ☐ constant

(5) Let A, B be sets and $A \times B$ the Cartesian product. By projection onto the second factor, one understands the map π_2 as:

☐ $\begin{aligned} A \times B &\longrightarrow A \\ (a,b) &\longmapsto b \end{aligned}$ ☐ $\begin{aligned} A \times B &\longrightarrow B \\ (a,b) &\longmapsto b \end{aligned}$ ☐ $\begin{aligned} B &\longrightarrow A \times B \\ b &\longmapsto (a,b) \end{aligned}$

(6) Let $f : X \to Y$ be a map. Which of the following statements implies that f is surjective?

☐ $f^{-1}(Y) = X$ ☐ $f(X) = Y$ ☐ $f^{-1}(X) = Y$

(7) Let $X \xrightarrow{f} Y \xrightarrow{g} Z$ be maps. Then the map $gf : X \to Z$ is defined by

☐ $x \longmapsto g(f(x))$ ☐ $x \longmapsto f(g(x))$ ☐ $x \longmapsto g(x)(f)$

(8) Let

$$\begin{array}{ccc} & \overset{f}{\nearrow} & Y \\ X & & \uparrow h \\ & \underset{g}{\searrow} & Z \end{array}$$

be a commutative diagram. Then we have

☐ $h = gf$ ☐ $f = hg$ ☐ $g = fh$

(9) The map $f : \mathbb{R}\smallsetminus\{0\} \to \mathbb{R}\smallsetminus\{0\}$, $x \mapsto \frac{1}{x}$ is bijective. The inverse map $f^{-1} : \mathbb{R}\smallsetminus\{0\} \to \mathbb{R}\smallsetminus\{0\}$ is defined by

☐ $x \longmapsto \frac{1}{x}$ ☐ $x \longmapsto x$ ☐ $x \longmapsto -\frac{1}{x}$

(10) $\mathbb{R} \to \mathbb{R}$, $x \mapsto x^2$ is

☐ surjective but not injective
☐ injective but not surjective
☐ neither surjective nor injective

1.4 Remarks on the Literature

I imagine that the reader of the beginning of a first-year text is just starting her or his studies and therefore might be interested in what a lecturer — in this case, I — thinks of the relation between books and lectures.

Many years ago, as I was preparing the notes for my students, from which this book has emerged, the books and manuscripts on linear algebra occupied four feet of shelf space in the departmental library; today, they occupy more than fifteen. According to mood one can find this reassuring or terrifying, but one thing has certainly not changed: a beginning student in mathematics actually needs no textbook. The lectures are autonomous, and the most important work for the student is her or his *own* lecture notes. This perhaps strikes you as a task from mediaeval times. Take notes? Somewhere in the fifteen feet there must be some book containing the material of the lectures. And if I don't have to write with the lecturer, then I can *think* much better with him — so you say. Besides, you say to yourself: write? And what if I can't decipher what is written on the board? Or what if the lecturer writes so fast that I can't follow him?[1] And what if I'm ill and can't come to the lecture? Then I'm stuck with my fragmentary notes.

[1] "I can't even *speak* as fast as Jänich writes" was one student's comment passed on to me.

So plausible appear these arguments, and yet they do not hold water. First of all, on average no book in those fifteen feet will contain the "material of the lectures." Indeed, the large number of books and manuscripts on linear algebra is more a sign that each lecturer prefers to go his own way. Of course, many a lecture course is rooted in a given book or manuscript, and then you should have the book, if only because the lecturer may leave gaps to be filled from it. But even then you should take notes, and as soon as he makes use of two books, you can be certain that he will follow neither of them exactly. If you can't write fast enough, you must train yourself to do so; if you can't read the board from the back of the room, you must look for a seat nearer the front; and if you are ill, you must copy a colleague's notes. Why this effort? If not, you will lose touch with the material, fall behind, and soon understand nothing else being taught. Ask any older student if he has ever learned anything in a lecture course in which he has not taken notes. It is as if information presented to the eye and ear must first pass through the hand in order really to enter the brain. Perhaps this is linked to the fact that in practicing mathematics, you again have to write. But whatever the reason, experience proves it.

When you are really in the swing of a course of lectures, books will be very useful to you, and for more senior year studies books are essential. You must therefore learn to work with books, but as a novice you must not lightly let a book tempt you away from direct contact with the course.

1.5 Exercises

1.1: If $f : X \to Y$ is a map, one calls the set $\{(x, f(x)) \mid x \in X\}$ the ***graph*** Γ_f of f. The graph is a subset of the Cartesian product $X \times Y$. In diagram (a) it is indicated by a line. But the graph of a map cannot be an arbitrary subset of $X \times Y$ since, for example, to each x there can only correspond *one* $f(x)$, and thus the line drawn in diagram (b) is not a graph. The exercise is to draw graphs of functions with preassigned properties. For example, the graph of a nonsurjective map is illustrated in (c).

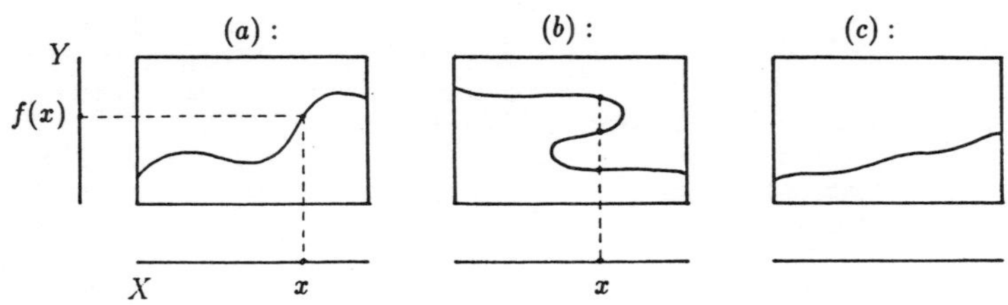

Fig. 12a. Graph   Fig. 12b. Nongraph   Fig. 12c. Graph of nonsurjective map

Give, in this fashion, examples of graphs of maps f with the following properties:

(i) f surjective, but not injective
(ii) f injective, but not surjective
(iii) f bijective
(iv) f constant
(v) f neither injective nor surjective
(vi) $X = Y$ and $f = \mathrm{Id}_X$
(vii) $f(X)$ consists of only two elements

1.2: The inverse map f^{-1} of a bijective map $f : X \to Y$ clearly has the properties $f \circ f^{-1} = \mathrm{Id}_Y$ and $f^{-1} \circ f = \mathrm{Id}_X$, since in the first case each element $f(x) \in Y$ is mapped by $f(x) \mapsto x \mapsto f(x)$ onto $f(x)$, and in the second case each $x \in X$ is mapped by $x \mapsto f(x) \mapsto x$ onto x. Conversely, one has the following (and the proof is the point of the exercise):

Let $f : X \to Y$ and $g : Y \to X$ be maps such that $fg = \mathrm{Id}_Y$ and $gf = \mathrm{Id}_X$, then f is bijective and $f^{-1} = g$.

(An injectivity proof runs like this: "Let $x, x' \in X$ and $f(x) = f(x')$, then ... . Therefore $x = x'$, and f is proved to be injective." On the other hand, the pattern for a surjectivity proof is: "Let $y \in Y$. Choose $x = \ldots$. Then we have ... , therefore $f(x) = y$, and f is proved to be surjective.")

1.3: Let

$$\begin{array}{ccc} X & \xrightarrow{\;f\;} & Y \\ {\scriptstyle\alpha}\uparrow{\scriptstyle\cong} & & {\scriptstyle\beta}\uparrow{\scriptstyle\cong} \\ A & \xrightarrow[\;g\;]{} & B \end{array}$$

be a commutative diagram of maps with α, β bijective. Show that g is injective if and only if f is injective.

(We shall frequently meet this kind of diagram in the text. The situation is then mostly: f is the object of our interest, α and β are subsidiary constructions, means to an end, and we already know something about g. This information about g then tells us something about f. In solving this exercise you will see into the mechanism of this information transfer.)

Chapter 2
Vector Spaces

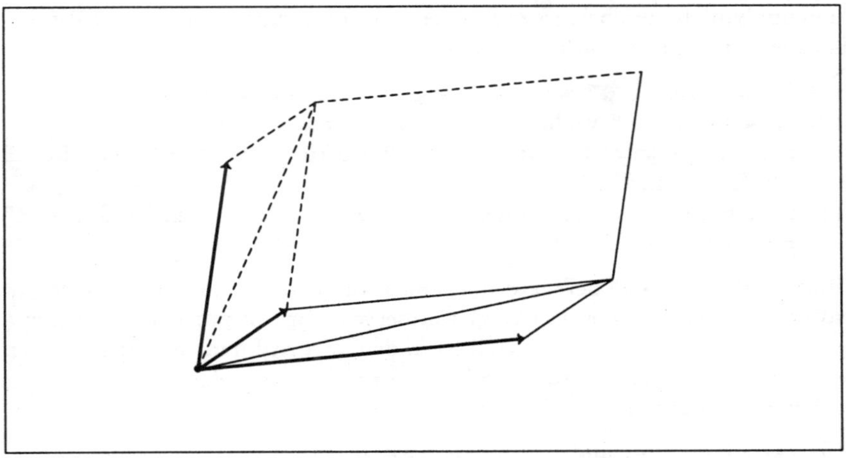

2.1 Real Vector Spaces

Vector spaces and not vectors are the main topic of linear algebra. The elements of a vector space are called vectors, and in order to explain the mathematical notion of a "vector," we first need the concept of a vector *space*. The individual properties of vectors are irrelevent; what matters is that addition and scalar multiplication in the vector space satisfy certain axioms or rules.

I will first illustrate these rules by means of an important example, indeed a model example of a real vector space, $\mathbb{R}^n$. The elements of this set are the n-tuples of real numbers, and with numbers one can *calculate* in various ways. Thus we can add n-tuples of real numbers together if we introduce the following definition.

Definition: Let $(x_1, \ldots, x_n)$ and $(y_1, \ldots, y_n)$ be n-tuples of real numbers. We define their sum by

$$(x_1, \ldots, x_n) + (y_1, \ldots, y_n) := (x_1 + y_1, \ldots, x_n + y_n) .$$

The sum is again an n-tuple of real numbers. In a similar way one can define what it means to multiply an n-tuple $(x_1, \ldots, x_n)$ by a real number λ.

Definition: Let $\lambda \in \mathbb{R}$ and $(x_1, \ldots, x_n) \in \mathbb{R}^n$. We define

$$\lambda(x_1, \ldots, x_n) := (\lambda x_1, \ldots, \lambda x_n) \in \mathbb{R}^n.$$

Since these computational operations amount simply to doing with each component of an n-tuple what we otherwise do with individual numbers, from the rules applying to numbers we derive corresponding rules applying to n-tuples. Thus with regard to *addition* we note:

(1) For all $x, y, z \in \mathbb{R}^n$ we have $(x + y) + z = x + (y + z)$.
(2) For all $x, y \in \mathbb{R}^n$ we have $x + y = y + x$.
(3) If for the sake of brevity we write 0 for $(0, \ldots, 0) \in \mathbb{R}^n$, then for all $x \in \mathbb{R}^n$ we have $x + 0 = x$.
(4) If we write $-(x_1, \ldots, x_n)$ instead of $(-x_1, \ldots, -x_n)$, then for all $x \in \mathbb{R}^n$ we have $x + (-x) = 0$.

(Hint on the notation: here x denotes an n-tuple of real numbers. We do not have enough letters in order forever to reserve x for this purpose. Ahead, in a quite different connection, we shall let x denote a real number lying between -1 and 1. But in each case we will say to which set x belongs.)

For *multiplication* by real numbers we have

(5) For all $\lambda, \mu \in \mathbb{R}$ and $x \in \mathbb{R}^n$ we have $\lambda(\mu x) = (\lambda\mu)x$.
(6) For all $x \in \mathbb{R}^n$ we have $1x = x$.

Finally, both "distributive laws" hold for the "compatibility" of *addition* and *multiplication*:

(7) For all $\lambda \in \mathbb{R}$ and $x, y \in \mathbb{R}^n$ we have $\lambda(x + y) = \lambda x + \lambda y$.
(8) For all $\lambda, \mu \in \mathbb{R}$ and $x \in \mathbb{R}^n$ we have $(\lambda + \mu)x = \lambda x + \mu x$.

This was our first example: a small excursion into calculating with n-tuples of real numbers. Now let us consider a quite different set, with whose elements one can also calculate.

One calls a map $X \to \mathbb{R}$ a real-valued *function* on X. Let M be the set of real valued functions on the interval $[-1, 1]$, that is, $M = \{f \mid f : [-1, 1] \to \mathbb{R}\}$. If $f, g \in M$ and $\lambda \in \mathbb{R}$ we define the functions $f + g$ and λf by $(f + g)(x) := f(x) + g(x)$ and $(\lambda f)(x) := \lambda f(x)$ for all $x \in [-1, 1]$. Then $f + g$ and λf are again elements of M.

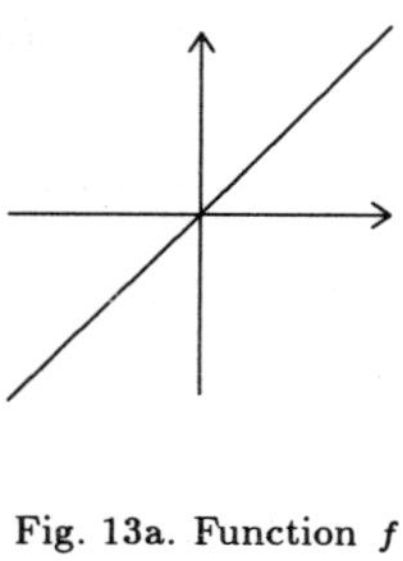

Fig. 13a. Function f

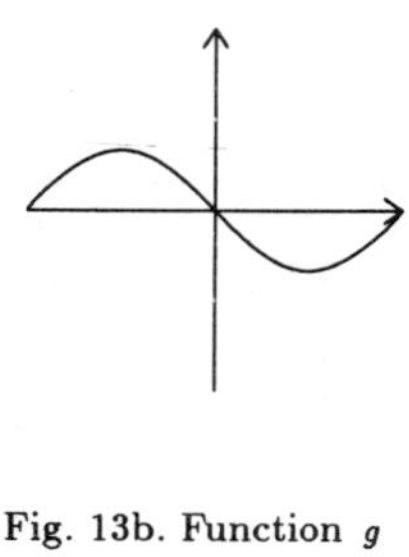

Fig. 13b. Function g

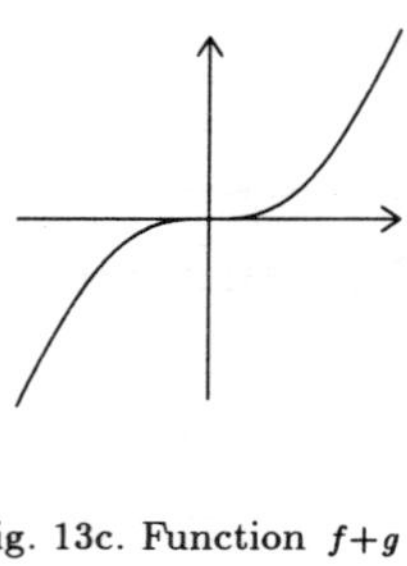

Fig. 13c. Function $f+g$

The eight rules we have just listed for $\mathbb{R}^n$ also hold for M: if we let 0 denote the element of M defined by $0(x) := 0$ for all $x \in [-1,1]$, and $-f$ the element defined by $(-f)(x) := -f(x)$, then for all $f, g, h \in M$, $\lambda, \mu \in \mathbb{R}$ we have

(1) $(f+g)+h = f+(g+h)$
(2) $f+g = g+f$
(3) $f+0 = f$
(4) $f+(-f) = 0$
(5) $\lambda(\mu f) = (\lambda\mu)f$
(6) $1f = f$
(7) $\lambda(f+g) = \lambda f + \lambda g$
(8) $(\lambda+\mu)f = \lambda f + \mu f$

So far as the eight rules go, these functions behave in the same way as n-tuples of real numbers, even though a single function considered on its own is quite different to an n-tuple.

Definition: A triple $(V, +, \cdot)$ consisting of a set V, a map (called addition)

$$\begin{aligned} + : V \times V &\longrightarrow V \\ (x,y) &\longmapsto x+y \end{aligned}$$

and a map (called scalar multiplication)

$$\begin{aligned} \cdot : \mathbb{R} \times V &\longrightarrow V \\ (\lambda, x) &\longmapsto \lambda x \cdot \end{aligned}$$

is called a ***real vector space*** if the following eight axioms hold for the maps $+$ and $\cdot$:

(1) $(x+y)+z = x+(y+z)$ for all $x, y, z \in V$.
(2) $x+y = y+x$ for all $x, y \in V$.
(3) There exists an element $0 \in V$ (called "zero" or the "zero vector") with $x+0 = x$ for all $x \in V$.
(4) For each element $x \in V$ there exists an element $-x \in V$ with $x+(-x) = 0$.
(5) $\lambda(\mu x) = (\lambda\mu)x$ for all $\lambda, \mu \in \mathbb{R}$, $x \in V$.
(6) $1x = x$ for all $x \in V$.
(7) $\lambda(x+y) = \lambda x + \lambda y$ for all $\lambda \in \mathbb{R}$, $x, y \in V$.
(8) $(\lambda+\mu)x = \lambda x + \mu x$ for all $\lambda, \mu \in \mathbb{R}$, $x \in V$.

I have already given you two examples: the space $(\mathbb{R}^n, +, \cdot)$ of n-tuples of real numbers, and the space $(M, +, \cdot)$ of real functions on the interval $[-1,1]$. And many more vector spaces have their place in mathematics. Let us talk about function spaces. In the example above the functions are defined on the interval $[-1,1]$, but this is unimportant for the vector space properties.

The set of all real functions on an *arbitrary* domain of definition D with the same addition and scalar multiplication becomes a vector space. But instead of considering *all* functions on D, it is usually more interesting to study functions on D with particular important properties. Thus there exist vector spaces of continuous functions, vector spaces of differentiable functions, vector spaces of solutions to homogeneous linear differential equations, and many more. One cannot anticipate which function spaces one will meet later on.

This is similar for the spaces of n-tuples: often it is not a matter of considering *all* n-tuples, but only the space of *those* n-tuples satisfying a particular system of homogeneous linear equations. Moreover, there exist many vector spaces, the elements of which are neither n-tuples nor functions. You will be meeting some of these soon — for example, vector spaces of matrices or vector spaces of endomorphisms or of operators, and later on others, such as the vector space of translations of an affine space, tangent spaces to surfaces and other manifolds, vector spaces of differential forms, and vector spaces, whose names as yet mean nothing to you, like real cohomology groups or Lie algebras. And this is only an enumeration of a few more or less *concrete* examples of vector spaces. Often one has to do with vector spaces, for which in addition to the axioms one has additional information (as for example with *Hilbert* or *Banach* spaces) not necessarily, however, involving knowledge of individual properties of the elements.

Because here we shall be doing linear algebra for the axiomatically defined vector space concept above, and not for $\mathbb{R}^n$ alone, it follows that already in the first weeks and months of your university career you will learn something essential about these various and in part difficult mathematical areas. This is really great! And indeed mathematics has taken a long time to attain this modern structural point of view. But you will perhaps suspect that we must pay a high price for this. Surely linear algebra is much harder for an abstract vector space than for $\mathbb{R}^n$? In no way, I reply — in many ways it is even simpler and clearer. But we do not obtain the great advantages *completely* free, and particularly at the start we must check certain statements for abstract vector spaces, which are obvious for spaces of n-tuples or of functions. It may appear strange and a little disturbing that such things have need of proof. The following remarks are examples of this. But don't worry; in Exercise 2.1 we will clear things away.

Remark 1: In a vector space there only exists one zero vector, for if 0 and $0'$ were zero vectors, we would have

$$0 = 0 + 0' = 0' + 0 = 0'$$

(by axioms 2 and 3).

Remark 2: In a vector space for each x there exists only one $-x$.

PROOF: If $x + a = 0$ and $x + b = 0$, we have

$$\begin{aligned} a &= a + 0 && \text{(axiom 3)} \\ &= a + (x + b) && \text{(by assumption)} \\ &= (a + x) + b && \text{(axiom 1)} \\ &= (x + a) + b && \text{(axiom 2)} \\ &= 0 + b && \text{(by assumption)} \\ &= b + 0 && \text{(axiom 2)} \\ &= b && \text{(axiom 3), and hence } \quad a = b. \end{aligned}$$

□

> **Notational convention:** From now on, as usual, we shall simply write $x - y$ instead of $x + (-y)$.

Before we pass to the next section (complex numbers and complex vector spaces), I would like to draw your attention to an important peculiarity of mathematical notation — namely, the frequent *double meaning* of symbols. For example, we have labeled the zero vector as 0. This does not mean that the real number zero, also written as 0, should be an element of the vector space, but instead that in V there exists exactly one vector, the addition of which "does nothing," which is called the zero vector, and which *like the number zero* is denoted by 0.

If we generally were to admit that, within a proof or within some other context, one and the same symbol could have distinct meanings, then soon our communication would break down. And each individual case of such double meaning is of course a possible source of confusion, particularly for beginners; this cannot be denied.

On the other hand, we must clearly take note of the fact that double meanings cannot be completely avoided; in fact, mathematical literature is full of them. Strict avoidance of double meanings would in the course of time lead to such an overload of symbols, that even quite simple statements would sink under their own weight. Because of its limited contents I could for a while avoid all double meanings in this work, but then I would have to adopt some very strange notational practices, which later on would give you difficulties in making the unavoidable switch to normal mathematical fare.

But we shall be as sparing as possible with double meanings, avoid cases with a real danger of confusion, and calmly name those cases which do occur. To label the zero vector with 0 is clearly such a case. It will be clear from the context whether number or vector is intended. For example, if $x, y \in V$ and $x + y = 0$, then this 0 is clearly a vector, etc. I would like at once to make you aware of another case of double meaning: in what follows, instead of the "vector space $(V, +, \cdot)$," we shall mostly say the "vector space V." This gives the symbol V a double meaning both as the vector space and as the set V underlying the vector space.

2.2 Complex Numbers and Complex Vector Spaces

In many mathematical questions working with real numbers only resembles studying point distributions on lines and finding no system:

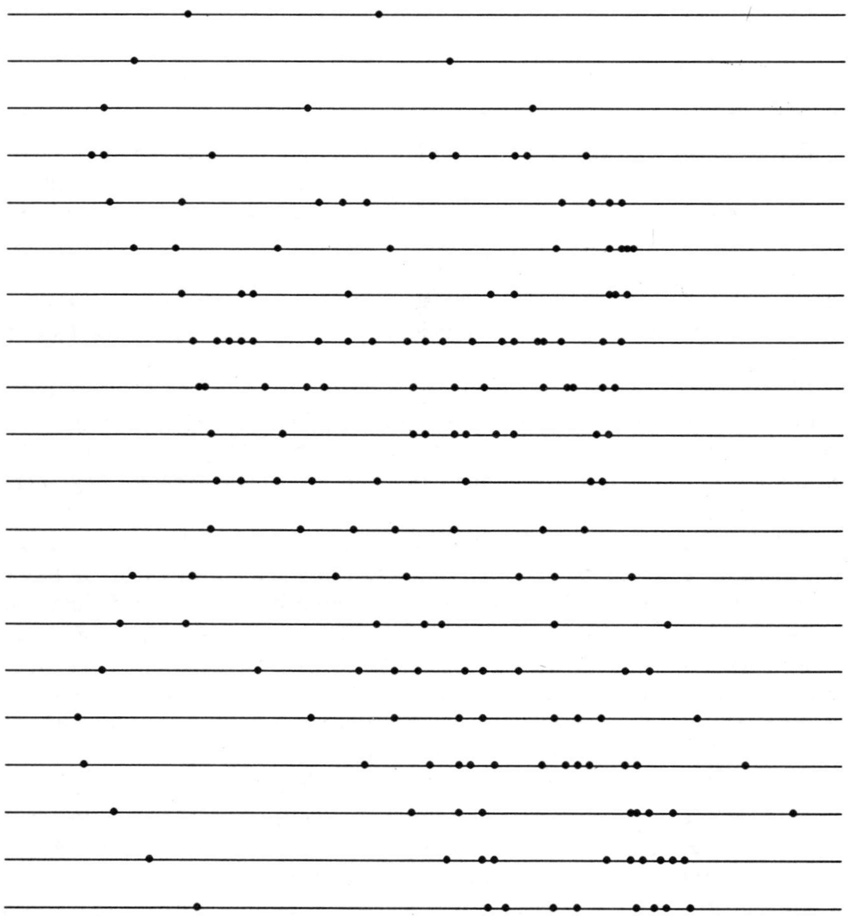

whereas working with *complex* numbers makes clear what is going on. The complex number system often makes possible decisive insights into the structure and methods of "real" mathematics.

Definition: By the ***field of complex numbers***, one understands the set $\mathbb{C} := \mathbb{R}^2$ together with the two operations

$$+ : \mathbb{C} \times \mathbb{C} \longrightarrow \mathbb{C} \qquad \text{("Addition") and}$$
$$\cdot : \mathbb{C} \times \mathbb{C} \longrightarrow \mathbb{C} \qquad \text{("Multiplication")},$$

which are defined by

$$(x, y) + (a, b) := (x + a, y + b) \quad \text{and}$$
$$(x, y) \cdot (a, b) := (xa - yb, xb + ya).$$

Thus, addition is the same as in the real vector space $\mathbb{R}^2$, but on first sight multiplication gives the impression of being arbitrary, and like one of those formulae, which experience shows one tends to forget. Why not simply define $(x, y)(a, b) = (xa, yb)$, which would seem to be the most obvious? This is best explained by first introducing an alternative notation for the elements of $\mathbb{R}^2$.

Notation: Let $\mathbb{R} \times 0 \subset \mathbb{C}$ play the role of $\mathbb{R}$, so that we write $x \in \mathbb{C}$ instead of $(x, 0) \in \mathbb{C}$, and in this way treat $\mathbb{R}$ as a subset of $\mathbb{C}$. In order to distinguish the elements of $0 \times \mathbb{R}$ we shorten $(0, 1)$ to i, so that now each $(0, y)$ can be written as yi and each (x, y) as $x + yi$:

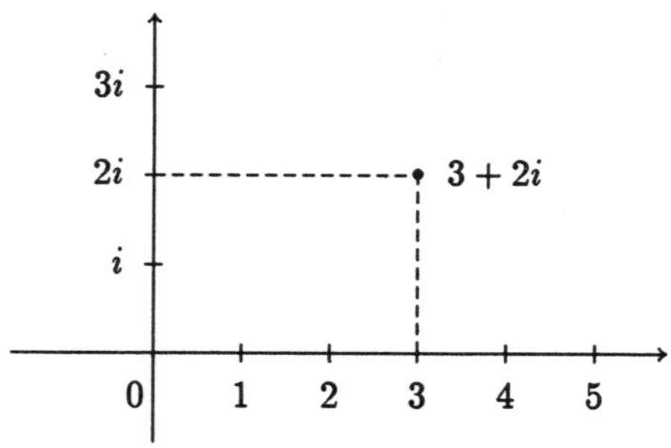

Multiplication in $\mathbb{C}$ should satisfy the following: first of all it should be associative, commutative, and distributive with respect to addition, that is, for all $u, v, w \in \mathbb{C}$, we should have $(uv)w = u(vw)$, $uv = vu$, and $u(v + w) = uv + uw$. All this would be satisfied by the multiplication $(x, y)(a, b) = (xa, yb)$. Next, multiplication by a real number x should be "scalar multiplication" in the real vector space $\mathbb{R}^2$, thus $x(a + bi) = xa + xbi$. (This is already *not* satisfied by $(x, y)(a, b) := (xa, yb)$.) Finally, and historically this was the real motive for introducing complex numbers, the so-called "imaginary numbers" yi were meant to serve as square roots for the *negative* real numbers, that is, their squares should be negative real numbers! One achieves this by setting $i^2 = -1$. Now if a multiplication in $\mathbb{C}$ with all these properties does exist at all, then necessarily $(x+yi)(a+bi) = xa+yia+xbi+yibi = xa-yb+(ya+xb)i$, and this is where the formula given in the definition for multiplication comes from.

There is more to say about the "inner mechanics" of complex multiplication (for example, multiplication by i is just rotation through 90°), but for our purposes in linear algebra it suffices for the time being to note that one calculates with complex numbers "exactly as" with real numbers. In particular, the following properties of complex multiplication are important for us.

Remark: Complex multiplication $\mathbb{C} \times \mathbb{C} \to \mathbb{C}$ is associative, commutative, and distributive; has a "unit"; and admits inversion of elements distinct from zero. Restricted to $\mathbb{R} \times \mathbb{C} \to \mathbb{C}$, it is the scalar multiplication in $\mathbb{R}^2$, and to $\mathbb{R} \times \mathbb{R} \to \mathbb{R} \subset \mathbb{C}$ the usual multiplication of real numbers.

Expressed as formulae the properties set out in the first sentence of the above remark say that: for all $u, v, w \in \mathbb{C}$ we have $(uv)w = u(vw)$, $uv = vu$, $u(v + w) = uv + uw$, $1u = u$, and if $u \neq 0$, there exists a unique $u^{-1} \in \mathbb{C}$ with $u^{-1}u = 1$.

Complex vector spaces are defined analogously to real vector spaces: one has only to replace $\mathbb{R}$ by $\mathbb{C}$ and "real" by "complex" in all instances.

Then $\mathbb{C}^n := \mathbb{C} \times \cdots \times \mathbb{C}$ is an example of a complex vector space, exactly as $\mathbb{R}^n$ is one of a real vector space. The first four axioms, which have to do only with addition in V, are taken over word for word. Perhaps it is better once more to write out the whole definition.

DEFINITION: A triple $(V, +, \cdot)$ consisting of a set V, a map $+ : V \times V \to V$, $(x, y) \mapsto x + y$, and a map $\cdot : \mathbb{C} \times V \to V$, $(\lambda, x) \mapsto \lambda x$, is called a ***complex vector space*** if the following axioms hold:

(1) For all $x, y, z \in V$ we have $(x + y) + z = x + (y + z)$.
(2) For all $x, y \in V$ we have $x + y = y + x$.
(3) There exists an element $0 \in V$ so that, for all $x \in V$ we have $x + 0 = x$.
(4) For each $x \in V$ there exists $-x \in V$ with $x + (-x)$.
(5) For all $\lambda, \mu \in \mathbb{C}$ and $x \in V$ we have $\lambda(\mu x) = (\lambda\mu)x$.
(6) For all $x \in V$ we have $1x = x$.
(7) For all $\lambda \in \mathbb{C}$, $x, y \in V$ we have $\lambda(x + y) = \lambda x + \lambda y$.
(8) For all $\lambda, \mu \in \mathbb{C}$, $x \in V$ we have $(\lambda + \mu)x = \lambda x + \mu x$.

Instead of "real vector space" one also says "vector space over $\mathbb{R}$," and instead of "complex vector space," we use "vector space over $\mathbb{C}$." If we speak of a "vector space over $\mathbb{F}$" in what follows, then we mean that $\mathbb{F}$ equals either $\mathbb{R}$ or $\mathbb{C}$.

2.3 Vector Subspaces

If V is a vector space over $\mathbb{F}$ and $U \subset V$ is a subset, then it is clear that one can add elements of U together and multiply by elements of $\mathbb{F}$, but this is far from making U into a vector space. For example, it can happen that $x + y \notin U$, even though $x, y \in U$ (see Fig. 14), and if this is so, then the addition in V gives no map $U \times U \to U$, as we would need for a vector space U, but only a map $U \times U \to V$. First of all then we must require, if U is to become a vector space through the V-addition and V-scalar multiplication, that for all $x, y \in U$ and $\lambda \in \mathbb{F}$ we have $x + y \in U$, $\lambda x \in U$. Moreover, we must make sure that $U \neq \varnothing$, since otherwise axiom 3 (existence of zero) cannot be satisfied. But this is then enough, and the validity of the axioms follows automatically. We will formulate this as a remark, but first we give the definition.

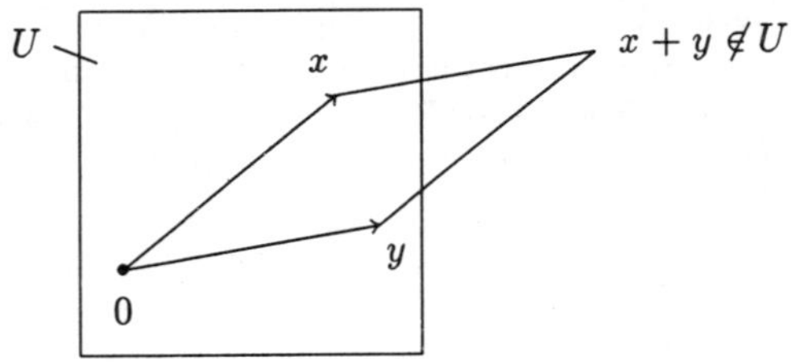

Fig. 14. Elements of $U \subset V$ can be added, but this does not necessarily define an addition "in U."

Definition: Let V be a vector space over $\mathbb{F}$. A subset $U \subset V$ is called a ***vector subspace*** (or just ***subspace*** for short) of V if $U \neq \varnothing$ and for all $x, y \in U$ and all $\lambda \in \mathbb{F}$ we have $x + y \in U$, $\lambda x \in U$.

Remark: If U is a subspace of V, then U contains the zero vector of V, and for each $x \in U$ the vector $-x \in V$ is contained in U also.

PROOF: One is tempted to say that this is obvious from $U \neq \varnothing$ and $\lambda x \in U$ for all $\lambda \in \mathbb{F}$, $x \in U$, since one can then put $\lambda = 0$ or $\lambda = -1$. For spaces of functions or n-tuples this is indeed clear, but since $(V, +, \cdot)$ is an arbitrary vector space, we must look for a *proof* that $0 \cdot x = 0$ and $(-1) \cdot x = -x$, since there is nothing in the axioms about this.

However, we have $0 \cdot x = (0 + 0) \cdot x = 0 \cdot x + 0 \cdot x$ by axioms 3 and 7, hence $0 = 0 \cdot x + (-0 \cdot x) = (0 \cdot x + 0 \cdot x) + (-0 \cdot x)$ by axiom 4. Therefore, $0 = 0 \cdot x + (0 \cdot x + (-0 \cdot x)) = 0 \cdot x + 0 = 0 \cdot x$ by 1 and 4, hence $0 \cdot x = 0$, as we wanted to show. As a consequence we also obtain the other assertion, since we now know that $0 = 0 \cdot x = (1 + (-1)) \cdot x = 1 \cdot x + (-1) \cdot x = x + (-1) \cdot x$. Thus $x + (-1) \cdot x = 0$, that is, $(-1) \cdot x = -x$. □

Going through the eight axioms for U, one sees the following.

Corollary: If U is a subspace of V, then U together with the addition and scalar multiplication inherited from $(V, +, \cdot)$ is itself a vector space over $\mathbb{F}$.

In particular, $\{0\}$ and V are themselves subspaces of V. In the pictorial representation of $\mathbb{R}^3$ as "space," the subspaces other than $\{0\}$ and $\mathbb{R}^3$ are the "planes" and "lines" passing through the origin.

That the *intersection* of subspaces of V is again a vector subspace of V is now too obvious as to deserve a formal proof. (Really?) But it is an important fact, so we include it below.

Fact: If U_1, U_2 are vector subspaces of V, then $U_1 \cap U_2$ is also a vector subspace of V.

2.4 Test

(1) Let $n \geq 1$. Then $\mathbb{R}^n$ consists of

- ☐ n real numbers
- ☐ n-tuples of real numbers
- ☐ n-tuples of vectors

(2) Which of the following statements is not an axiom for real vector spaces?

- ☐ For all $x, y \in V$ we have $x + y = y + x$.
- ☐ For all $x, y, z \in V$ we have $(x + y) + z = x + (y + z)$.
- ☐ For all $x, y, z \in V$ we have $(xy)z = x(yz)$.

(3) For the multiplication of complex numbers we have $(x + yi)(a + bi) =$

- ☐ $xa + ybi$
- ☐ $xy + yb + (xb - ya)i$
- ☐ $xa - yb + (xb + ya)i$

(4) In a vector space V over $\mathbb{F}$ scalar multiplication is given by a map

- ☐ $V \times V \to \mathbb{F}$
- ☐ $\mathbb{F} \times V \to V$
- ☐ $\mathbb{F} \times \mathbb{F} \to \mathbb{F}$

(5) Which formulation below can be completed correctly to give the definition of the concept of a real vector space?

- ☐ A set V is called a real vector space if there exist two maps $+ : \mathbb{R} \times V \to V$, and $\cdot : \mathbb{R} \times V \to V$, so that the following eight axioms are satisfied ...

- ☐ A set of real vectors is called a real vector space, if the following eight axioms are satisfied ...
- ☐ A triple $(V, +, \cdot)$ in which V is a set and $+$ and $\cdot$ are respectively maps $V \times V \to V$ and $\mathbb{R} \times V \to V$ is called a real vector space if the following eight axioms are satisfied ...

(6) Which of the following statements is true? If V is a vector space over $\mathbb{F}$ then

- ☐ $\{x + y \mid x \in V, y \in V\} = V$.
- ☐ $\{x + y \mid x \in V, y \in V\} = V \times V$.
- ☐ $\{\lambda v \mid \lambda \in \mathbb{F}, v \in V\} = \mathbb{F} \times V$.

(7) Which of the following statements is true?

- ☐ If U is a subspace of V, then $V \setminus U$ is also a subspace of V.
- ☐ There exists a subspace U of V for which $V \setminus U$ is also a subspace, but $V \setminus U$ is not a subspace for all subspaces U.
- ☐ If U is a subspace of V, then $V \setminus U$ is never a subspace of V.

(8) Which of the following subsets $U \subset \mathbb{R}^n$ is a vector subspace?

- ☐ $U = \{x \in \mathbb{R}^n \mid x_1 = \cdots = x_n\}$
- ☐ $U = \{x \in \mathbb{R}^n \mid x_1^2 = x_2^2\}$
- ☐ $U = \{x \in \mathbb{R}^n \mid x_1 = 1\}$

(9) On restriction of scalar multiplication to the scalar domain $\mathbb{R}$, a complex vector space $(V, +, \cdot)$ becomes a real vector space $(V, +, \cdot|\mathbb{R} \times V)$. In particular, $V := \mathbb{C}$ can itself be regarded as a real vector space in this way. Do the imaginary numbers $U = \{iy \in \mathbb{C} \mid y \in \mathbb{R}\}$ then form a vector subspace?

- ☐ Yes, because then $U = \mathbb{C}$.
- ☐ Yes, because $0 \in U$ and when $\lambda \in \mathbb{R}$ and $ix, iy \in U$, we also have $i(x + y) \in U$ and $i\lambda x \in U$.
- ☐ No, because λiy does not need to be imaginary, since for example $i^2 = -1$.

(10) How many vector subspaces does $\mathbb{R}^2$ have?

- ☐ two: $\{0\}$ and $\mathbb{R}^2$
- ☐ four: $\{0\}$, $\mathbb{R} \times 0$, $0 \times \mathbb{R}$ (the "axes") and $\mathbb{R}^2$ itself
- ☐ infinitely many

2.5 Fields

A section for mathematicians

Besides $\mathbb{R}$ and $\mathbb{C}$ there exist many other so-called "fields" that one can use as scalar domains for vector spaces.

Definition: A ***field*** is a triple $(\mathbb{F}, +, \cdot)$ consisting of a set $\mathbb{F}$ and two rules of composition

$$\begin{aligned} + : \mathbb{F} \times \mathbb{F} &\longrightarrow \mathbb{F} \\ (\lambda, \mu) &\longrightarrow \lambda + \mu \qquad (\text{"Addition"}) \end{aligned}$$

and

$$\begin{aligned} \cdot : \mathbb{F} \times \mathbb{F} &\longrightarrow \mathbb{F} \\ (\lambda, \mu) &\longrightarrow \lambda\mu \qquad (\text{"Multiplication"}) \end{aligned}$$

which satisfy the following axioms:

(1) For all $\lambda, \mu, \nu \in \mathbb{F}$ we have $(\lambda + \mu) + \nu = \lambda + (\mu + \nu)$.
(2) For all $\lambda, \mu \in \mathbb{F}$ we have $\lambda + \mu = \mu + \lambda$.
(3) There exists an element $0 \in \mathbb{F}$ with $\lambda + 0 = \lambda$ for all $\lambda \in \mathbb{F}$.
(4) For each $\lambda \in \mathbb{F}$ there exists an element $-\lambda \in \mathbb{F}$ with $\lambda + (-\lambda) = 0$.
(5) For all $\lambda, \mu, \nu \in \mathbb{F}$ we have $(\lambda\mu)\nu = \lambda(\mu\nu)$.
(6) For all $\lambda, \mu \in \mathbb{F}$ we have $\lambda\mu = \mu\lambda$.
(7) There exists an element $1 \in \mathbb{F}$, $1 \neq 0$, such that we have $1\lambda = \lambda$ for all $\lambda \in \mathbb{F}$.
(8) For all $\lambda \in \mathbb{F}$ with $\lambda \neq 0$ there exists $\lambda^{-1} \in \mathbb{F}$ with $\lambda^{-1}\lambda = 1$.
(9) For all $\lambda, \mu, \nu \in \mathbb{F}$ we have $\lambda(\mu + \nu) = \lambda\mu + \lambda\nu$.

These nine properties are of course modeled on calculation with real or complex numbers, and as a first approximation one can say that one calculates in a field "just as" one calculates in $\mathbb{R}$ or $\mathbb{C}$.

One can deduce easily from the axioms that the elements 0 and 1 named in axiom 3 and axiom 7 are uniquely determined, so that we can speak of "the zero" and "the one" (or "the unit") of the field. Furthermore $-\lambda$ and λ^{-1} are uniquely determined for a given λ; we have that $(-1)\lambda = -\lambda$, that $\lambda\mu = 0$ if and only if $\lambda = 0$ or $\mu = 0$, and that $(-1)(-1) = 1$. We note this for the convenience of the reader of the main text.

Fact: 0 and 1 are uniquely determined, likewise $-\lambda$ and λ^{-1} for given λ. We have $(-1)\lambda = -\lambda$, $(-1)(-1) = 1$ and $\lambda\mu = 0$ if and only if $\lambda = 0$ or $\mu = 0$.

If $\mathbb{F}$ is an arbitrary field, one defines the concept of a ***vector space over*** $\mathbb{F}$ analogously to real vector spaces — replace $\mathbb{R}$ by $\mathbb{F}$ everywhere. In this book, when we speak of vector spaces over $\mathbb{F}$, then for the reader of the *Sections for Mathematicians* we mean that $\mathbb{F}$ is an arbitrary field, unless otherwise specified. In particular what we have formulated above for "vector fields over $\mathbb{F}$" holds for arbitrary fields and not only, as was first given, for $\mathbb{F} = \mathbb{R}$ and $\mathbb{F} = \mathbb{C}$.

Let me formulate the definition of a "field" in yet another way, which in my view is easier to remember. This definition needs a little preliminary discussion, and therefore I have not used it in the main text. Thus: if anywhere in mathematics you see an operation denoted by "+" (and this happens frequently), then you can be reasonably certain that this operation is *associative* and *commutative*, that is, for all x, y, z for which the operation is defined, we have (1) $x + (y + z) = (x + y) + z$ and (2) $x + y = y + x$. If in addition there exists a "neutral element" 0, and each x has its negative, then one calls the set together with the operation $+$ an *abelian group* (after the Norwegian mathematician Niels Henrik Abel (1802–1829)).

DEFINITION: An ***abelian group*** is a pair $(A, +)$ consisting of a set A and an operation $+ : A \times A \to A$, such that we have

(1) $(a + b) + c = a + (b + c)$ for all $a, b, c \in A$.
(2) $a + b = b + a$ for all $a, b \in A$.
(3) There exists an element $0 \in A$ with $a + 0 = a$ for all $a \in A$.
(4) For each $a \in A$ there exists an element $-a \in A$ with $a + (-a) = 0$.

The zero is then uniquely determined as is $-a$ for a. The standard example for an abelian group is $\mathbb{Z}$, the abelian group of integers.

In principle it does not matter which symbol one uses for the operation; if the four axioms are satisfied, we have an abelian group. However, two forms of notation have become standard: first, the "additive" notation used in the above definition. In the second, the "multiplicative" notation, one writes the operation as $\cdot : G \times G \to G$, $(g, h) \mapsto gh$, calls the neutral element 1 rather than 0, and the "negative" g^{-1} rather than $-g$. The definition as such remains the same, so $(G, \cdot)$ is called a (multiplicative) abelian group, if

(1) $(gh)k = g(hk)$ for all $g, h, k \in G$.
(2) $gh = hg$ for all $g, h \in G$.
(3) There exists $1 \in G$ with $1g = g$ for all $g \in G$.
(4) For each $g \in G$ there exists $g^{-1} \in G$ with $g^{-1}g = 1$.

With the notation introduced in this way, one can reformulate the definition of a field as follows.

Note: $(\mathbb{F}, +, \cdot)$ is a field if and only if $(\mathbb{F}, +)$ and $(\mathbb{F} \setminus 0, \cdot)$ are abelian groups, and the operations are distributive in the usual sense; thus $\lambda(\mu + \nu) = \lambda\mu + \lambda\nu$ and $(\mu + \nu)\lambda = \mu\lambda + \nu\lambda$ for all $\lambda, \mu, \nu \in \mathbb{F}$. □

In the analogy between field axioms and the properties of addition and multiplication for real numbers, one must keep an eye on one danger with the field calculations. This is linked with the double meaning of 1 as a number and as a field element. Thus, for the multiplicatively neutral element of a field, one uses the notation 1, and one also denotes $1+1 \in \mathbb{F}$ by 2, etc. In this way each symbol for a natural number has a double meaning as both a number and a field element, and in the same way, $n\lambda$ also has a double meaning for each $\lambda \in \mathbb{F}$. Taking n to be a natural number $n\lambda := \lambda + \lambda + \cdots + \lambda$ (n summands) has to do with field *addition* only, and one uses the same notation in an arbitrary additively written abelian group. If, on the other hand, one takes n to be the field element, then $n\lambda$ has a meaning as a product in the sense of field *multiplication*. Luckily this makes no difference, since for $1, 2, \ldots \in \mathbb{F}$ we have $\lambda + \lambda = 1\lambda + 1\lambda = (1+1)\lambda = 2\lambda$, etc., because of axiom 9, and therefore in both interpretations $n\lambda$ is the same field element. However, the element $n\lambda$ can be zero, even though neither the *number* n nor the field element λ is zero. It can indeed happen that

$$1 + \cdots + 1 = 0 \in \mathbb{F}$$

for an appropriate number of summands!

Definition: Let $\mathbb{F}$ be a field and $1 \in \mathbb{F}$ its unit element. For positive natural numbers n we write $n1 := 1 + \cdots + 1 \in \mathbb{F}$ (n summands). If $n1 \neq 0$ for all $n > 0$ we say that $\mathbb{F}$ is a ***field of characteristic zero***. Otherwise the ***characteristic*** of $\mathbb{F}$ is defined to be the smallest positive number p such that $p1 = 0$.

Remark: If the characteristic p of $\mathbb{F}$ is not zero, then it is a prime number.

PROOF: Since $1 \neq 0$ (axiom 7) $p = 1$ is not possible. If now $p = p_1 p_2$ with $p_1 > 1$, $p_2 > 1$, then we would have $(p_1 p_2)1 = (p_1 1)(p_2 1) = 0$, hence either $p_1 1 = 0$ or $p_2 1 = 0$, contradicting the fact that $p_1 p_2$ is the smallest positive number n such that $n1 = 0$. □

Examples: The fields $\mathbb{R}$, $\mathbb{C}$, and $\mathbb{Q}$ (the field of rational real numbers) all have characteristic 0. If p is a prime number, then one can make $\{0, 1, \ldots, p-1\}$ into a field $\mathbb{F}_p$ by defining the sum and the product to be the remainders of the usual sum and product on division by p. (Example: $3 \cdot 4 = 12$ in $\mathbb{Z}$, $12 : 7 = 1$ with remainder 5, so $3 \cdot 4 = 5$ in $\mathbb{F}_7$.) The field $\mathbb{F}_p$ has characteristic p. In particular, for $\mathbb{F}_2 = \{0, 1\}$ define addition and multiplication by $0 + 0 = 0$, $1 + 0 = 0 + 1 = 1$, and $1 + 1 = 0$, and $0 \cdot 0 = 0 \cdot 1 = 1 \cdot 0 = 0$, and $1 \cdot 1 = 1$, respectively, thus making $\mathbb{F}_2$ into a field of characteristic 2.

2.6 What Are Vectors?

A section for physicists

From the mathematical standpoint this question has a simple but satisfactory answer: vectors are the elements of a vector space. However, as a physicist you are confronted with another point of view, when, for example, in the Berkeley Physics Course [4], page 25, you read that "a vector is a quantity having direction as well as magnitude." What is meant here? What does this have to do with the mathematical concept of vectors? Is it the same, only in different words? Very legitimate questions, but not so easy to answer. The concept is not completely the same, anyway.

For a closer explanation, I must examine the three words *quantity, direction,* and *magnitude.* As a preparation let me begin with what one understands in mathematics by the magnitude of a vector. Then we will return to our problem and try to build a bridge between the mathematical and physical concepts of a vector.

In the mathematical sense, vectors initially have no "magnitude," but we can give them a magnitude. Whether and how we do this depends on our reasons for considering the vector space at hand in the first place. In those parts of the first seven chapters of this book that are not specially directed at physicists, we have no reason to equip vectors with a magnitude, because the mathematical questions considered have nothing to do with magnitudes. So for the moment let us jump across five chapters, right into Chapter 8, where the *Euclidean vector spaces*, real vector spaces with an inner product, are considered.

By the *magnitude* or *norm* or *length* of a vector $x \in \mathbb{R}^n$, one understands the number

$$\|x\| := \sqrt{x_1^2 + \cdots + x_n^2} \in \mathbb{R}.$$

The reason for making this definition is given by Pythagoras's theorem from elementary geometry, depicted in Fig. 15.

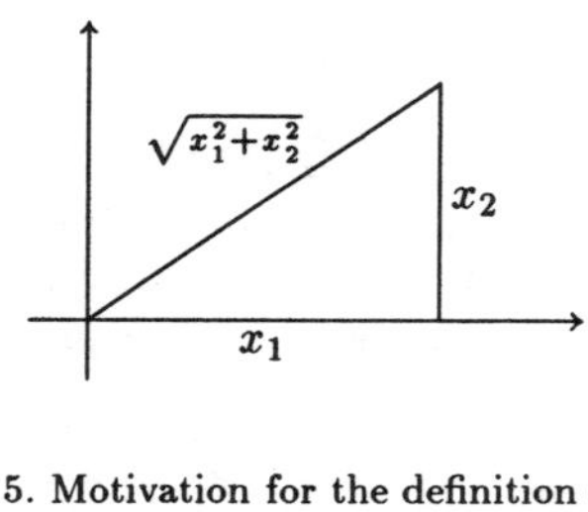

Fig. 15. Motivation for the definition of $\|x\|$

A vector $e \in \mathbb{R}^n$ is then called a unit vector if we have $\|e\| = 1$. If $x \neq 0$, for example, $e := \frac{x}{\|x\|}$ is a unit vector.

For two vectors $x, y \in \mathbb{R}^n$ one calls the number

$$\langle x, y\rangle := x_1 y_1 + \cdots + x_n y_n \in \mathbb{R}$$

the standard ***inner product*** of x and y.

We have $\langle x, x\rangle = \|x\|^2$ and also $\|x+y\|^2 - \|x-y\|^2 = 4\langle x, y\rangle$, so that, interpreted in the elementary geometric sense, $\langle x, y\rangle = 0$ means that the diagonals $x+y$ and $x-y$ have equal length in the parallelogram generated by x and y. This means that this is a rectangle, and that x and y are perpendicular to each other: $\langle x, y\rangle = 0 \iff x \perp y$. From this follows the elementary geometric significance of $\langle x, y\rangle$ for arbitrary vectors $x, y \in \mathbb{R}^n$ distinct from zero: if we let $e := \frac{x}{\|x\|}$ and $e' := \frac{y}{\|y\|}$, and λe is the foot of the perpendicular from e' on to the line $0x$, and if we let $\alpha(x, y)$ denote the angle between x and y, so that as a consequence $\lambda = \cos\alpha(x, y)$, then e is perpendicular to $e' - \lambda e$. Therefore $\langle e, e'\rangle - \lambda\langle e, e\rangle = 0$, or $\langle e, e'\rangle = \lambda$. But this means that

$$\frac{\langle x, y\rangle}{\|x\|\|y\|} = \cos\alpha(x, y),$$

so that the inner product describes not only lengths but also elementary angle measures in $\mathbb{R}^n$ (see Fig. 16). In mathematics many other real vector spaces besides $\mathbb{R}^n$ are studied, and therefore one introduces a general concept, which imitates the most important properties of the standard inner product in $\mathbb{R}^n$, as the following definition explains.

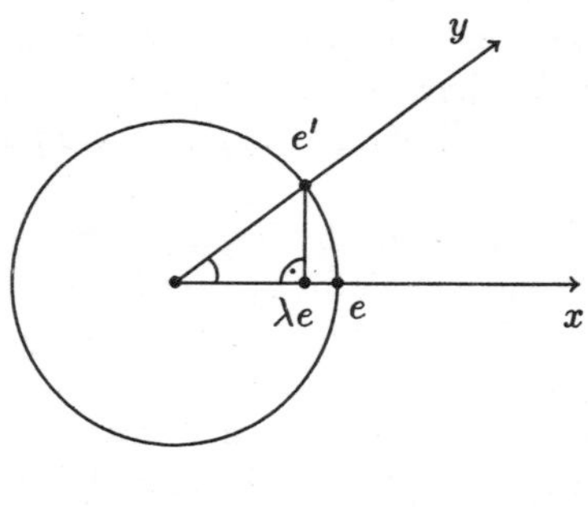

Fig. 16. Determining the angle between two vectors x and y from their inner product

Definition: Let V be a real vector space. By an ***inner product on*** V one understands a map $V \times V \to \mathbb{R}$ that is *bilinear, symmetric*, and *positive definite*, that is, if the map is written as $(v, w) \mapsto \langle v, w\rangle$, we have

(1) For each $w \in V$ the map $\langle\,\cdot\,, w\rangle : V \to \mathbb{R}$ is *linear*, which means that we always have $\langle v_1 + v_2, w\rangle = \langle v_1, w\rangle + \langle v_2, w\rangle$ and $\langle \lambda v, w\rangle = \lambda\langle v, w\rangle$, and analogously for fixed v and $\langle v, \cdot\,\rangle : V \to \mathbb{R}$ (*bilinearity*).
(2) $\langle v, w\rangle = \langle w, v\rangle$ for all v, w (*symmetry*).
(3) $\langle v, v\rangle \geq 0$ for all v, and $\langle v, v\rangle = 0$ only for $v = 0$ (*positive definiteness*).

For each real vector space V one can find such an inner product $\langle\ ,\ \rangle$, indeed there are many of them — on $\mathbb{R}^n$ there also exist inner products besides the standard one — and in order to fix lengths of vectors and angles between them one has first to choose an inner product.

Definition: A ***Euclidean vector space*** is a pair $(V, \langle\ ,\ \rangle)$ consisting of a real vector space V and an inner product $\langle\ ,\ \rangle$ on it.

Or one can also say that the vector space V is *given* or comes *equipped with* an inner product.

Definition: In a Euclidean vector space, $\|v\| := \sqrt{\langle v, v\rangle}$ is called the ***magnitude*** or ***length*** of v, and for $v \neq 0$, $w \neq 0$,

$$\alpha(v,w) := \arccos \frac{\langle v,w\rangle}{\|v\|\|w\|},$$

is called the ***angle*** between v and w.

So much for the magnitude of a vector from the *mathematical* point of view, and I return to the problem, into which I have allowed myself to be led, of analyzing the difference between the mathematical and physical concepts of a vector.

This difference is connected with the fact that in physics there exists an ever-present space of overwhelming importance, while linear algebra refuses it "official recognition" as a mathematical object. This space is the *actual physical space* in which we all find ourselves.

Unofficially, of course, this *observation space* is well known to mathematicians. But if we are to take the points in this space to be "definite objects of our intuition or of our thought," our mathematical conscience will feel rather uneasy. In the present discussion, however, we must ignore such philosophical niceties and boldly declare the observation space to be sufficiently well defined.

This does not make the observation space $\mathcal{A}$ into a vector space — where, for example, is its zero? But it is closely related to certain vector spaces. Namely, if one chooses some arbitrary point $O \in \mathcal{A}$, and calls this the zero or origin, then one can describe all points $P \in \mathcal{A}$ as so-called *position vectors* $\overrightarrow{OP}$ *with respect to* O. These are illustrated by arrows from O to P, can be multiplied by real numbers, and added together as in the "parallelogram of forces" shown in Fig. 17,

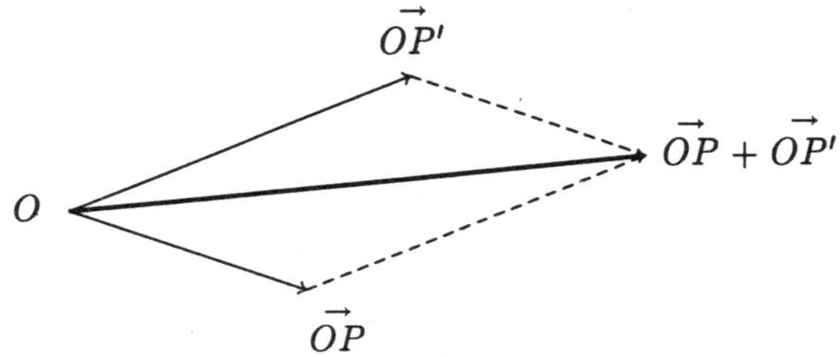

Fig. 17. Definition of addition $\mathcal{A}_O \times \mathcal{A}_O \to \mathcal{A}_O$

thus defining a vector space, which we now denote by $\mathcal{A}_O$. Morover, at any fixed point O in space, physics does not consider position vectors only, but also electric field strength vectors, velocity vectors, force vectors and many others. By means of superposition and so forth, addition and scalar multiplication are physically defined, making electric field strength vectors at the point O into a real vector space $\mathcal{E}_O$, velocity vectors into a space $\mathcal{V}_O$, force vectors into a space $\mathcal{F}_O$, and so on.

Although it is not customary in physics to give special names and notation to the vector spaces containing such vectors as elements, I have invented these names here in order to use them in our comparison of the mathematical and physical concepts of a vector. So far we see nothing problematic in this comparison: we merely have a few concrete physical examples, $\mathcal{A}_O$, $\mathcal{E}_O$, $\mathcal{V}_O$, $\mathcal{F}_O$, etc., of the general mathematical concept of a real vector space in front of us. Things become more interesting, however, when we consider the physically measured *magnitude* of the physical vectors.

With every position vector $\vec{r} \in \mathcal{A}_O$, with every electric vector $\vec{E} \in \mathcal{E}_O$, etc., is associated a (physical) magnitude $|\vec{r}|$, respectively $|\vec{E}|$, etc. In general this physical magnitude is not just a *number* but also has a *physical dimension.* (This use of the word *dimension* has nothing to do with the concept of dimension in linear algebra, of which we will speak in the next section.) So, for example, we do not have that $|\vec{r}| = 5$ or $|\vec{E}| = 5$, but $|\vec{r}| = 5\,\mathrm{cm}$ and $|\vec{E}| = 5\,\frac{\mathrm{volts}}{\mathrm{cm}}$. Of course, a length-valued magnitude could easily be turned into one which is real-valued; we need "only" to decide on a unit of measure. But this is just what we want to avoid, since calculations with physical vectors should not depend on such an arbitrary choice. Instead of building our bridge by eliminating physical dimensions from physical formulae, we prefer to *introduce them* into linear algebra! We do this in the following way.

We consider a *length-valued scalar domain*, as one can perhaps call it, namely,

$$\mathbb{R}[\mathrm{length}] := \mathbb{R}[\mathrm{cm}] := \{x\,\mathrm{cm} \mid x \in \mathbb{R}\}.$$

In an obvious way this is a vector space; the operations are given by $x\,\mathrm{cm} + y\,\mathrm{cm} = (x + y)\mathrm{cm}$ and $\lambda \cdot (x\,\mathrm{cm}) = \lambda x\,\mathrm{cm}$. In the mathematical sense it is as "one-dimensional" as $\mathbb{R}$ itself. This length-valued scalar domain is not just a formal construction, but is physically interpretable as the vector space — not exactly of lengths, since what does a negative length mean? — but of length *differences*. Note that by this construction of the length-valued scalar domain we have in no way fixed the choice of a unit of measurement, since $\mathbb{R}[\mathrm{cm}]$ is precisely the same as $\mathbb{R}[\mathrm{m}]$, just as $5\,\mathrm{cm}$ is precisely the same as $0.05\,\mathrm{m}$. Thus, for position vectors we have $|\vec{r}| \in \mathbb{R}[\mathrm{cm}]$.

We do the same thing with all other physical dimensions; we have an electric field strength scalar domain $\mathbb{R}[\frac{\mathrm{volts}}{\mathrm{cm}}]$, a velocity scalar domain $\mathbb{R}[\frac{\mathrm{cm}}{\mathrm{sec}}]$, and so forth, all independent of the choice of units. We also want to have the dimensionless scalar domain $\mathbb{R}$ itself as an example; we can write it as $\mathbb{R} = \mathbb{R}[1]$. One can multiply these physical scalars together, for example, $5\,\mathrm{cm} \cdot 6\,\frac{\mathrm{volt}}{\mathrm{cm}} = 30\,\mathrm{volt} \in \mathbb{R}[\mathrm{volt}]$. You will not insist on long formal explanations of $\mathrm{cm} \cdot \mathrm{sec} = \mathrm{sec} \cdot \mathrm{cm}$ and $\frac{\mathrm{cm}}{\mathrm{cm}} = 1$.

The essential point that distinguishes the physical from the mathematical concept of a vector concerns the relation of the different physical vector spaces $\mathcal{A}_O$, $\mathcal{E}_O$, $\mathcal{V}_O$, ... *between each other.* As we have seen, each of these real vector spaces has its own physical scalar domain, to which the magnitudes of its vectors belong. But a physical vector space at the point O is also *characterized* by its scalar domain: there do not exist several kinds of electrical field strength vectors at the point O, distinguished somehow by the manner of their generation, but a vector at O with magnitude in $\mathbb{R}$[volts/cm] is an element of $\mathcal{E}_O$. Moreover, in physics it makes sense to multiply physical vectors by physical scalars; one then obtains vectors with appropriately altered scalar domain. For example, if one multiplies a velocity vector $\vec{v} \in \mathcal{V}_O$ by a time interval, say by $5\,\text{sec} \in \mathbb{R}[\text{sec}]$, one obtains a position vector $\vec{r} = 5\,\text{sec} \cdot \vec{v} \in \mathcal{A}_O$. Adopting our notation for scalar domains for this kind of products, we would here have, for example, $\mathcal{V}_O[\text{sec}] = \mathcal{A}_O$. In general, if $\mathcal{X}_O$ is a physical vector space at O with scalar domain $\mathbb{R}[a]$ and if $\mathbb{R}[b]$ is another scalar domain, then we want to write

$$\mathcal{X}_O[b] := \{b \cdot \vec{v} \mid \vec{v} \in \mathcal{X}_O\}.$$

This is then the physical vector space with scalar domain $\mathbb{R}[ab]$. Note that here again we have not met any specification of a unit of measure, since different units of measure are only distinguished by a real nonzero factor. In this way all physical vector spaces at O are canonically related to each other. If $\mathcal{X}_O$ and $\mathcal{Y}_O$ have scalar domains $\mathbb{R}[a]$ and $\mathbb{R}[b]$, we then have $\mathcal{Y}_O = \mathcal{X}_O[\frac{b}{a}]$ and $\mathcal{X}_O = \mathcal{Y}_O[\frac{a}{b}]$. In particular, if we so wish, we can describe them all by means of the space of position vectors: $\mathcal{X}_O = \mathcal{A}_O[\frac{a}{\text{cm}}]$. *Each physical vector at the point O is a position vector up to multiplication by some positive physical scalar factor.* In this way it is possible to say that a physical vector, even if it is not a position vector, has a "direction": namely a direction in space!

If we look once more from our present position on the formulation: *a vector is a quantity having direction as well as magnitude*, we can clearly make the distinction with the mathematical concept of a vector.

(1) A physical vector is a *quantity* with some physical origin. However general this may be, it already expresses some other interpretation of a vector, since the mathematical vector space axioms make no requirement as to the origin or characteristic of the vectors.

(2) A physical vector has a *magnitude*, which is not part of the initial description of a mathematical vector. If, however, one introduces magnitudes by means of the additional structure of an inner product, these are real numbers, and not, as in physics, dimension-laden physical scalars.

(3) Finally, a physical vector has a *direction* in (physical) space, because the physical vector spaces described above have a close relation to the position vector space. There is no correspondence here with the mathematical concept of a vector, since the axioms make no mention of a physical space.

As is to be expected, these differences have further consequences. In the Euclidean vector spaces of linear algebra, for example, the (real-valued) magnitude is defined by the (equally real-valued) inner product: $\|x\| = \sqrt{\langle x, x\rangle}$, and conversely the inner product can be reconstructed from the magnitude: $4\langle x, y\rangle = \|x + y\|^2 - \|x - y\|^2$. In physics the inner product $\vec{v} \cdot \vec{w}$ is formed in parallel from the physical vectors $\vec{v}$ and $\vec{w}$. But in contrast to the inner product in linear algebra, $\vec{v} \cdot \vec{w}$ is in general not a real number, but a physical scalar. Moreover, $\vec{v}$ and $\vec{w}$ do not need to belong to the same physical vector space; for example, we are allowed to multiply position vectors $\vec{r} \in \mathcal{A}_O$ by electrical vectors $\vec{E} \in \mathcal{E}_O$ and obtain $\vec{r} \cdot \vec{E} \in \mathbb{R}[\text{volts}]$. To what extent are assertions about the mathematical inner product still applicable in physics? For example, it would be difficult to write $4\vec{r} \cdot \vec{E} = |\vec{r} + \vec{E}|^2 - |\vec{r} - \vec{E}|^2$, since the sum of a position and an electrical vector makes no sense.

Perhaps in the telling of this tale you begin to wonder whether, as a physicist, you are sitting in the right lecture hall. Shouldn't linear algebra, insofar as it is relevant for physicists, teach the basics of calculation with *physical* vectors?

Not exactly. The real use of linear algebra for the working physicist is that linear algebra, although elementary in itself, provides an essential tool for more advanced branches of *mathematics*, which are indispensable for physics. Thus, differential and integral calculus for several variables, and the theory of differential equations, have considerable need of the linear algebra dealing with "mathematical" vector spaces, without mentioning the mathematical methods of quantum mechanics or modern theoretical physics. For *this* reason you are learning linear algebra, and not for a better understanding of the parallelogram of forces. But this does not mean that "mathematical" linear algebra is not applicable to calculations with physical vectors — even where these are concerned with inner products. Let me finally say a few words on this.

Among the physical vector spaces at the point O is a particularly peculiar one, namely the (physically) *dimensionless* vector space, for which we make the notation

$$\mathcal{U}_O := \mathcal{A}_O[\tfrac{1}{\text{cm}}] = \mathcal{E}_O[\tfrac{\text{cm}}{\text{volt}}] = \cdots \text{ etc.}$$

For vectors in this vector space, magnitude is actually real-valued. For example, if $\vec{r} \in \mathcal{A}_O$ is a position vector with a magnitude of $5\,\text{cm} \in \mathbb{R}[\text{cm}]$, then the dimensionless vector $\vec{u} := \frac{2}{\text{cm}} \cdot \vec{r}$ has the magnitude $|\vec{u}| = 10 \in \mathbb{R}$. This has nothing to do with the choice of unit of length. Thus, this physically determined, real-valued magnitude $\mathcal{U}_O \to \mathbb{R}$ fixes a genuinely mathematical inner product, that is, if one defines

$$\vec{u} \cdot \vec{v} := \frac{1}{4}(|\vec{u} + \vec{v}|^2 - |\vec{u} - \vec{v}|^2),$$

the resulting map

$$\begin{aligned} \mathcal{U}_O \times \mathcal{U}_O &\longrightarrow \mathbb{R} \\ (\vec{u}, \vec{v}) &\longmapsto \vec{u} \cdot \vec{v} \end{aligned}$$

is indeed bilinear, symmetric, and positive definite. In this way $\mathcal{U}_O$ is made into an Euclidean vector space according to the precise meaning of the mathematical definition. In the last analysis this depends on the fact that the Euclidean axioms hold in the observation space: the mathematical theory of Euclidean vector spaces applies directly to the physical vector space $\mathcal{U}_O$ without ifs and buts, without choice of units of measure or indeed coordinates. Because of the connections between the physical vector spaces, one has nevertheless entry to all other inner products between physical vectors. Namely, $\mathcal{A}_O = \mathcal{U}_O[\text{cm}]$, $\mathcal{E}_O = \mathcal{U}_O[\frac{\text{volt}}{\text{cm}}]$, etc., and using the inner product on $\mathcal{U}_O$ one obtains a composition (still called inner product)

$$\mathcal{U}_O[a] \times \mathcal{U}_O[b] \longrightarrow \mathbb{R}[ab],$$

for example, $\mathcal{A}_O \times \mathcal{E}_O \to \mathbb{R}[\text{volt}]$, $(\vec{r}, \vec{E}) \mapsto \vec{r} \cdot \vec{E}$, independently of the choice of units of measure. The Euclidean vector space $\mathcal{U}_O$ thus becomes a bridge between linear algebra, in which the real-valued inner product is always formed from two vectors in the same vector space, and calculations with vectors in physics, in which vectors of different kinds are inner-multiplied together with their product lying in some physical scalar domain.

For practical calculations with physical vectors it is often useful to introduce coordinates. In mathematical linear algebra one also uses coordinates in some, say, *real* vector space V. For this one first chooses a ***basis*** (see Chapter 3) of vectors $v_1, \ldots, v_n$ of V, which means that each $v \in V$ can be expressed in a unique way as $v = \lambda_1 v_1 + \cdots + \lambda_n v_n$ by means of real numbers $\lambda_1, \ldots, \lambda_n$. The $\lambda_1, \ldots, \lambda_n$ are then called ***coordinates*** of v, and the lines $g_i := \{\lambda v_i \mid \lambda \in \mathbb{R}\}$, $i = 1, 2, \ldots, n$, the ***coordinate axes*** of the space.

In calculations with physical vectors this is a little different, and again the dimensionless physical vector space $\mathcal{U}_O$ serves as a go-between. Thus, for calculation with physical vectors one takes a basis of $\mathcal{U}_O$, usually three mutually perpendicular unit vectors $\hat{x}, \hat{y}, \hat{z} \in \mathcal{U}_O$, so that

$$|\hat{x}| = |\hat{y}| = |\hat{z}| = 1 \quad \text{and}$$
$$\hat{x} \cdot \hat{y} = \hat{y} \cdot \hat{z} = \hat{z} \cdot \hat{y} = 0 .$$

How can we get such vectors? Well, for example, in the position vector space one chooses three mutually perpendicular (and nonzero) vectors $\vec{R}_x$, $\vec{R}_y$, $\vec{R}_z \in \mathcal{A}_O$ and lets

$$\hat{x} := \frac{\vec{R}_x}{|\vec{R}_x|}, \quad \hat{y} := \frac{\vec{R}_y}{|\vec{R}_y|} \quad \text{and} \quad \hat{z} := \frac{\vec{R}_z}{|\vec{R}_z|}.$$

Because $\vec{R}_x \in \mathcal{A}_O$ and $|\vec{R}_x| \in \mathbb{R}[\text{cm}]$, we have $\hat{x} \in \mathcal{U}_O$ and so forth. An electrical vector $\vec{E}_x \in \mathcal{E}_O$ in the direction $\vec{R}_x$ does the same service for us, again $\hat{x} = \vec{E}_x / |\vec{E}_x|$.

If then $\vec{v} \in \mathcal{U}_O[a]$ is some physical vector with magnitude in the scalar domain $\mathbb{R}[a]$, it can be written uniquely in the form

$$\vec{v} = v_x \hat{x} + v_y \hat{y} + v_z \hat{z}$$

with coordinates $v_x, v_y, v_z \in \mathbb{R}[a]$, and when one inner-multiplies both sides by $\hat{x}$, one obtains

$$\vec{v} \cdot \hat{x} = v_x,$$

and analogously $\vec{v} \cdot \hat{y} = v_y$ and $\vec{v} \cdot \hat{z} = v_z$. For example, if $\vec{E} \in \mathcal{E}_O$ is an electrical vector, one has

$$\vec{E} = E_x\hat{x} + E_y\hat{y} + E_z\hat{z}$$

with

$$E_x = \vec{E} \cdot \hat{x} \in \mathbb{R}[\tfrac{\text{volt}}{\text{cm}}],$$

and analogously for E_y and E_z. For the inner product of two physical vectors $\vec{v} \in \mathcal{U}_O[a]$ and $\vec{w} \in \mathcal{U}_O[b]$ using bilinearity and $\hat{x} \cdot \hat{x} = \hat{y} \cdot \hat{y} = \hat{z} \cdot \hat{z} = 1$, $\hat{x} \cdot \hat{y} = \hat{y} \cdot \hat{z} = \hat{z} \cdot \hat{x} = 0$, one immediately calculates that

$$\vec{v} \cdot \vec{w} = v_x w_x + v_y w_y + v_z w_z \in \mathbb{R}[ab],$$

giving, for example,

$$\vec{r} \cdot \vec{E} = r_x E_x + r_y E_y + r_z E_z \in \mathbb{R}[\text{volt}]$$

for the inner product of a position vector $\vec{r} \in \mathcal{A}_O$ with an electrical vector $\vec{E} \in \mathcal{E}_O$. This is the first of the *useful vector identities* on page 44 in the Berkeley Physics Course [4]. The others involve the vector product, which we will first consider in the next section. Note once more that introducing coordinates does not presuppose introducing units of measure.

While I am talking about these foundational questions, the section gets longer and longer, and I should take care that my book does not become lopsided. Nonetheless I *must* go into the question of what two physical vector spaces at two distinct points O and O' have to do with each other.

Is $\mathcal{E}_O$ the same vector space as $\mathcal{E}_{O'}$? No. One can look at it from either the mathematical or the physical point of view: a vector at the point O is not the same as a vector at the point O'.

However, we can use translation of the observation space, to move each position vector, and hence any other physical vector at the point O, to a corresponding vector at the point O'.

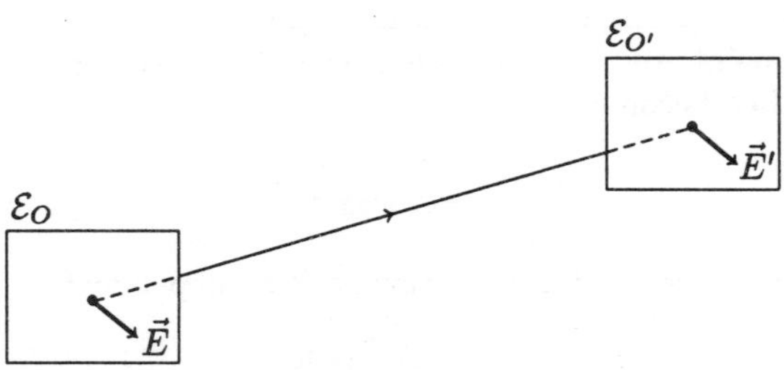

Fig. 18. Physical vector spaces $\mathcal{E}_O$ and $\mathcal{E}_{O'}$ related by translation

One can also say that $\vec{E}$ and $\vec{E}'$ represent the same *free* vector. According to one's taste one can either accept the concept of a "free" vector as hereby defined, or make use of a formal construction, such as taking a free physical vector to consist of all translates of a given "bound" physical vector. The free *electrical* vectors then form a vector space $\mathcal{E}_{\text{free}}$, and analogously we have $\mathcal{A}_{\text{free}}$, $\mathcal{U}_{\text{free}}$, etc. It only lacks a choice of origin to turn a free physical vector into a proper physical vector.

Why does one need free vectors? In physics, for example, it is often not a single vector but a whole vector *field* that is of interest. An electric field in some domain $\mathcal{B} \subset \mathcal{A}$ in space associates an electrical vector from $\mathcal{E}_O$ to each point $O \in \mathcal{B}$ (see Fig. 19).

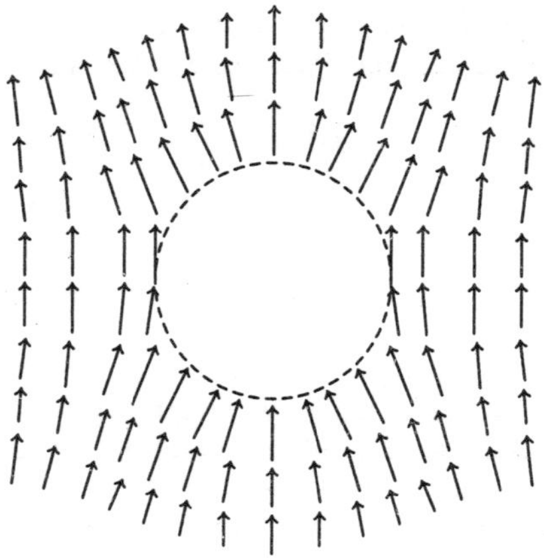

Fig. 19. Example of a vector field (electric field outside a dielectric sphere)

We can describe such a field by means of a map

$$\vec{E} : \mathcal{B} \to \mathcal{E}_{\text{free}} .$$

Instead of several vector spaces $\mathcal{E}_O$, $O \in \mathcal{B}$, we then have to deal only with one, which is more convenient. That we can be content with the free vector $\vec{E}(O) \in \mathcal{E}_{\text{free}}$ is not because it had suddenly become unimportant where the vector is situated, but because we know it: at O.

One can also use translation to move the dimensionless unit vectors $\hat{x}, \hat{y}, \hat{z}$ to an arbitrary point, and hence consider them as free vectors. An electric field $\vec{E} : \mathcal{B} \to \mathcal{E}_{\text{free}}$ then becomes

$$\vec{E} = E_x\hat{x} + E_y\hat{y} + E_z\hat{z},$$

where, however, the coordinates are now position dependent:

$$E_x : \mathcal{B} \to \mathbb{R}[\tfrac{\text{volt}}{\text{cm}}],$$

and similarly for y and z.

2.7 Complex Numbers 400 Years Ago

Historical aside

Mathematicians were first seriously confronted with complex numbers in the sixteenth century, when trying to solve equations. The simplest equations where one encounters "roots of negative numbers" are quadratic equations. However, it was not quadratic, but *cubic,* equations that forced the use of complex numbers, and this for good reason. By way of example let us consider the quadratic equation $x^2 + 3 = 2x$. The solution formula for this type of equation was known long before the sixteenth century. In our example, it would say $x = 1 + \sqrt{-2}$, which is a meaningless expression, since one cannot extract the roots of -2. The lack of sense of the solution formula could not disturb contemporary mathematicians in the least, since it corresponds to the fact that the equation actually has no solution (see Fig. 20). The thought that one might be able to *artificially enlarge* the number domain so that previously unsolvable equations would acquire solutions, and one would have a uniform theory of quadratic equations, is thoroughly modern and historically did *not* provide the pretext for discovering complex numbers. But the situation appears quite different, if one considers, say, the cubic equation $x^3 = 15x + 4$. In the sixteenth century one also found a solution formula for such equations, which in this case reads

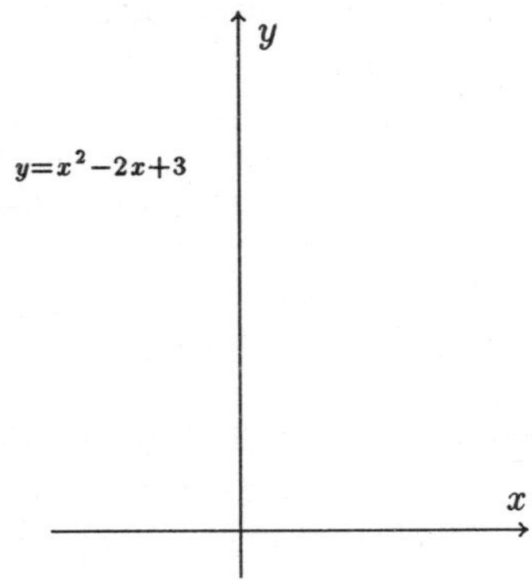

Fig. 20. $x^2-2x+3=0$ indeed has no solution.

$$x = \sqrt[3]{2 + \sqrt{-121}} + \sqrt[3]{2 - \sqrt{-121}}.$$

Again we have a meaningless expression, but this time it does correspond to the existence of the real solution $x = 4$. Sure, a root of -121 does not exist, but if one were to proceed as if it did, and if in calculating with this "imaginary number" one observes certain computational rules, then one can actually show that $\sqrt[3]{2+\sqrt{-121}} + \sqrt[3]{2-\sqrt{-121}} = 4$! The Italian engineer Rafael Bombelli actually performed systematic calculations with complex numbers in this way around 1560. However, mathematicians could not have been happy at first with these "imaginary numbers." On the one hand, they could not treat them as a mere game, since with their help mathematicians were able to obtain "genuine" (real) solutions of equations. On the other hand, they didn't "exist," and not all mathematicians accepted the use of such "devices." For a long time something mysterious clung to the complex numbers — Leibniz expressed that complex numbers were a kind of amphibian between being and not being. They were finally demystified in 1837 by the Irish mathematician and physicist

W. R. Hamilton, who was the first to treat complex numbers as they are handled today — by giving rules for the manipulation of pairs of real numbers.

(My source for this "historical aside": Helmuth Gericke, "History of the Concept of Number" (in German), see [2] in our list of references.)

2.8 Remarks on the Literature

It is not easy for students new to a subject to use books, because each book has its own system of notation, and there are small but irritating differences also in the definitions. One does one's best to arrive at a uniform terminology, but in a subject like linear algebra, which is used in almost all areas of mathematics, notational differences can hardly be avoided. When one considers that linear algebra is used in subjects as different as the numerical solution of systems of equations, homological algebra, and differential topology, for example, one has to be thankful that so much unity is still there!

"Getting into" a book parallel to the lectures requires patience, pen, paper, as well as confidence in the quality of the book. You can certainly have this confidence with P. R. Halmos, *Finite-Dimensional Vector Spaces* (no. [3] in our bibliography). Halmos is famous for his excellent expositional style: understandable, not dry and yet terse. Just try it! Our Chapter 2 covers §§ 1–4 and § 10 in Halmos — altogether just seven pages. Read these seven pages in order to get acquainted with the book. Our notation agrees well with that of Halmos. There are small differences: for example, instead of $\mathcal{Q}$, $\mathcal{R}$, $\mathcal{C}$, $\mathcal{Z}$, we write $\mathbb{Q}$, $\mathbb{R}$, $\mathbb{C}$, $\mathbb{Z}$; instead of "linear manifold" we say "subspace."

For the audience of my original lectures I added one special encouragement that my present readers will not need, namely to accustom oneself as soon as possible to reading textbooks written in English.

2.9 Exercises

Exercises for mathematicians

2.1: The rule $x+(y-x)=y$, for example, does not appear among the axioms of a vector space, but can be easily deduced from them:

$$\begin{aligned} x+(y-x) &= x+(-x+y) && \text{(by axiom 2)} \\ &= (x-x)+y && \text{(axiom 1)} \\ &= 0+y && \text{(axiom 4)} \\ &= y+0 && \text{(axiom 2)} \\ &= y && \text{(axiom 3).} \end{aligned}$$

So it would have been quite unnecessary to include $x+(y-x)=y$ among the axioms. In setting up an axiom system, one tries to choose the *fewest* and *simplest* axioms needed in order to deduce from them all the additional rules that one wants. Any use of such an additional rule must of course be justified by a proof that the rule actually is a logical consequence of the axioms.

But this does not mean that for each page of linear algebra you are bound to write ten pages of "reduction to the axioms." Given a little practice it can be assumed that you would be able to carry out reduction of your calculations to the axioms, and this does not need to be formally written out. You will gain this practice in the present exercise.

Let V be a vector space over $\mathbb{F} = \mathbb{R}$ or $\mathbb{F} = \mathbb{C}$, then for all $x \in V$ and all $\lambda \in \mathbb{F}$, prove

$$
\begin{array}{ll}
\text{(a)} & 0 + x = x \\
\text{(b)} & -0 = 0 \\
\text{(c)} & \lambda 0 = 0 \\
\text{(d)} & 0x = 0 \\
\text{(e)} & \lambda x = 0 \Longleftrightarrow \lambda = 0 \quad \text{or} \quad x = 0 \\
\text{(f)} & -x = (-1)x
\end{array}
$$

In fact (a) through (f) hold for vector spaces over arbitrary fields. Working out the proof in this generality is a useful exercise in the field axioms.

2.2: For $\alpha \in \mathbb{F}$ we define $U_\alpha := \{(x_1, x_2, x_3) \in \mathbb{F}^3 \mid x_1 + x_2 + x_3 = \alpha\}$. Show that U_α is a vector subspace of $\mathbb{F}^3$ if and only if $\alpha = 0$.

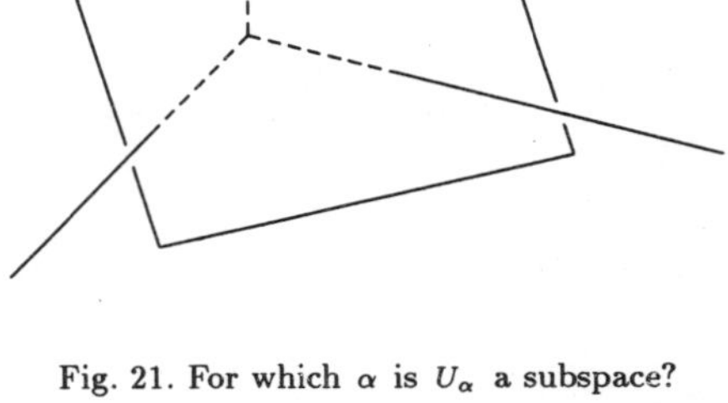

Fig. 21. For which α is U_α a subspace?

2.3: Let V be a vector space over $\mathbb{F}$ and U_1, U_2 be subspaces of V. Show that if $U_1 \cup U_2 = V$, then $U_1 = V$ or $U_2 = V$, or both. (This is a particularly pretty exercise. One can reduce the proof to three lines.)

The $*$-exercise

The $*$-exercises have a higher level of difficulty, but they do not require a higher level of knowledge than the regular exercises.

2*: If all the field axioms for $(\mathbb{F}, +, \cdot)$ (see the definition in Section 2.5) hold, with the possible exception of (8), then one calls $(\mathbb{F}, +, \cdot)$ a "commutative ring with unit." If in addition $\lambda\mu = 0$ only occurs if $\lambda = 0$ or $\mu = 0$, then $\mathbb{F}$ is called a "commutative ring with unit element and no divisors of zero" or an "integral domain" for short. Prove that every finite integral domain is a field.

Exercises for physicists

2.1P: Let (e_1, e_2, e_3) be an orthonormal basis of a three-dimensional Euclidean vector space. Let x, y be vectors with $x = 3e_1 + 4e_2$, $\|y\| = 5$, and $\langle y, e_3 \rangle \neq 0$. With these data calculate the cosine of the angle between $x + y$ and $x - y$. What can go wrong if $\langle y, e_3 \rangle = 0$?

2.2P: Let (e_1, e_2) be an *orthonormal basis* in a two-dimensional Euclidean vector space, that is, $\|e_1\| = \|e_2\| = 1$, $\langle e_1, e_2 \rangle = 0$, and all vectors of V have the form $\lambda_1 e_1 + \lambda_2 e_2$. Let $x = e_1 + e_2$. Show that $V_\alpha := \{v \in V \mid \langle v, x \rangle = \alpha\}$, $\alpha \in \mathbb{R}$, is a subspace of V if and only if $\alpha = 0$. Make a sketch to illustrate the positions of e_1, e_2, and V_1.

2.3P: See Exercice 2.3 for mathematicians.

CHAPTER 3

Dimension

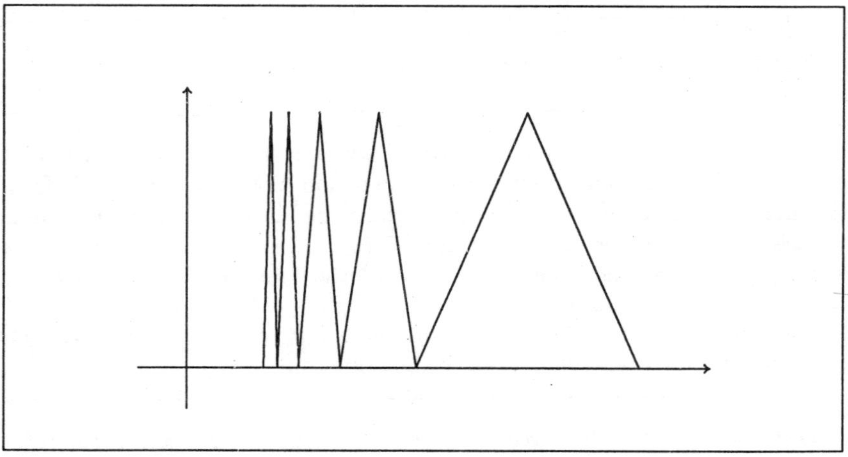

3.1 Linear Independence

Let V be a vector space over $\mathbb{F}$, let $v_1, \ldots, v_r$ be vectors, and let $\lambda_1, \ldots, \lambda_r$ be scalars, which means $v_i \in V$ and $\lambda_i \in \mathbb{F}$ for $i = 1, \ldots, r$. Then the vector $\lambda_1 v_1 + \cdots + \lambda_r v_r$ is called a *linear combination* of $v_1, \ldots, v_r$.

Definition: Let $v_1, \ldots, v_r \in V$. The set

$$L(v_1, \ldots, v_r) := \{\lambda_1 v_1 + \cdots + \lambda_r v_r \mid \lambda_i \in \mathbb{F}\} \subset V$$

of all ***linear combinations*** of $v_1, \ldots, v_r$ is called the ***linear hull*** of the r-tuple $(v_1, \ldots, v_r)$ of vectors. For the "0-tuple" consisting of no vectors, and denoted by $\varnothing$, we write $L(\varnothing) := \{0\}$.

Convention thus says that the zero vector can be constructed by linear combination "from nothing." If in what follows we talk of r-tuples of vectors, the 0-tuple $\varnothing$ is also considered admissible.

Since the sum of two linear combinations of $v_1, \ldots, v_r$ is again a linear combination of $v_1, \ldots, v_r$, because

$$(\lambda_1 v_1 + \cdots + \lambda_r v_r) + (\mu_1 v_1 + \cdots + \mu_r v_r) = (\lambda_1 + \mu_1)v_1 + \cdots + (\lambda_r + \mu_r)v_r,$$

and since, in addition, for each $\lambda \in \mathbb{F}$ the λ-multiple of a linear combination of $v_1, \dots, v_r$ is still such a linear combination, because

$$\lambda(\lambda_1 v_1 + \cdots + \lambda_r v_r) = (\lambda\lambda_1)v_r + \cdots + (\lambda\lambda_r)v_r,$$

and since, finally, $L(v_1, \dots, v_r)$ is not empty, $L(v_1, \dots, v_r)$ is a *subspace* of V. We record this as follows:

Fact: $L(v_1, \dots, v_r)$ is a subspace of V.

An r-tuple $(v_1, \dots, v_r)$ of elements of a vector space V is called *linearly dependent* if one of these vectors is a linear combination of the others. So far as the linear hull goes, one can omit this vector: the linear hull of the remaining vectors is the same as the linear hull of $(v_1, \dots, v_r)$. If $(v_1, \dots, v_r)$ is *not* linearly dependent, then one says that it is *linearly independent.* For practical use of the concept of linear independence, a slightly different, more "technical" formulation is in order. However, we will immediately convince ourselves that these two formulations come down to the same thing.

Definition: Let V be a vector space over $\mathbb{F}$. An r-tuple $(v_1, \dots, v_r)$ of vectors in V is said to be ***linearly independent*** if a linear combination of $(v_1, \dots, v_r)$ can only vanish if all the "coefficients" vanish; that is, if from $\lambda_1 v_1 + \cdots + \lambda_r v_r = 0$ it necessarily follows that $\lambda_1 = \cdots = \lambda_r = 0$. The 0-tuple $\varnothing$ is linearly independent.

Remark 1: $(v_1, \dots, v_r)$ is linearly independent if and only if none of these vectors is a linear combination of the others.

PROOF: We have two things to prove:

(a) $(v_1, \dots, v_r)$ linearly independent $\Longrightarrow$ no v_i is a linear combination of the others.

(b) no v_i is a linear combination of the others $\Longrightarrow$ $(v_1, \dots, v_r)$ is linearly independent.

For (a) suppose that $(v_1, \dots, v_r)$ is linearly independent. Assume that there exists some i with $v_i = \lambda_1 v_1 + \cdots + \lambda_{i-1}v_{i-1} + \lambda_{i+1}v_{i+1} + \cdots + \lambda_r v_r$. (This is a generally accepted way of indicating the omission of the ith term in the sum, even though it is not quite accurate say for $i = 1$.) But then the linear combination

$$\lambda_1 v_1 + \cdots + \lambda_{i-1}v_{i-1} + (-1)v_i + \lambda_{i+1}v_{i+1} + \cdots + \lambda_r v_r$$

would also be zero, even though not all coefficients are zero, since $-1 \neq 0$. This contradicts the linear independence of $(v_1, \dots, v_r)$, proving (a).

For (b) suppose that none of the v_i is a linear combination of the remaining vectors in $(v_1, \dots, v_r)$. Assume that $(v_1, \dots, v_r)$ is *linearly dependent.*

Then there exist $\lambda_1, \ldots, \lambda_r \in \mathbb{F}$ with $\lambda_i \neq 0$ for at least one i and $\lambda_1 v_1 + \cdots + \lambda_r v_r = 0$. However, it then follows that

$$v_i = -\frac{\lambda_1}{\lambda_i} v_1 - \cdots - \frac{\lambda_{i-1}}{\lambda_i} v_{i-1} - \frac{\lambda_{i+1}}{\lambda_i} v_{i+1} - \cdots - \frac{\lambda_r}{\lambda_i} v_r,$$

so v_i is a linear combination of the remaining vectors. This is a contradiction, and so (b) is also proved. □

Definition: Let V be a vector space over $\mathbb{F}$. An n-tuple $(v_1, \ldots, v_n)$ of vectors in V is called a ***basis*** of V, if the vectors are linearly independent and $L(v_1, \ldots, v_n) = V$.

If $(v_1, \ldots, v_n)$ is a basis, one can write each element $v \in V$ as a linear combination $v = \lambda_1 v_1 + \cdots + \lambda_n v_n$, and thus "generate" or "span" (as one says) the whole vector space by means of the vectors $v_1, \ldots, v_n$. But this already follows from $L(v_1, \ldots, v_n) = V$, so why does one require in addition that $v_1, \ldots, v_n$ be linearly independent? This condition has the effect that each $v \in V$ can be written in exactly *one* way as $\lambda_1 v_1 + \cdots + \lambda_n v_n$.

Remark 2: If $(v_1, \ldots, v_n)$ is a basis of V, then for each $v \in V$ there exists exactly one $(\lambda_1, \ldots, \lambda_n) \in \mathbb{F}^n$ with $v = \lambda_1 v_1 + \cdots + \lambda_n v_n$.

PROOF: Since $L(v_1, \ldots, v_n) = V$, for each $v \in V$ there exists some such $(\lambda_1, \ldots, \lambda_n) \in \mathbb{F}^n$. Suppose that $(\lambda'_1, \ldots, \lambda'_n) \in \mathbb{F}^n$ is another, hence

$$v = \lambda_1 v_1 + \cdots + \lambda_n v_n = \lambda'_1 v_1 + \cdots + \lambda'_n v_n .$$

Then $(\lambda_1 - \lambda'_1)v_1 + \cdots + (\lambda_n - \lambda'_n)v_n = v - v = 0$, and because of linear independence of $(v_1, \ldots, v_n)$ it follows that $\lambda_i - \lambda'_i = 0$, therefore $\lambda_i = \lambda'_i$ for $i = 1, \ldots, n$. □

In a certain sense one can say that one knows a vector space if one knows a basis of it. The $\mathbb{R}^n$ is not a good example since we "know" it in any case, but for the description of vector *subspaces*, for example, solution spaces for systems of equations, giving a basis is often the best method — we will return to this in Chapter 7, Systems of Linear Equations. More generally speaking, we will need bases in order to translate problems of linear algebra into the language of *matrices*.

The simplest concrete example of a basis of a vector space over $\mathbb{F}$ is the so-called ***canonical basis*** $(e_1, \ldots, e_n)$ of $\mathbb{F}^n$, where

$$\begin{aligned} e_1 &:= (1, 0, \ldots, 0) \\ e_2 &:= (0, 1, \ldots, 0) \\ &\vdots \\ e_n &:= (0, \ldots, 0, 1). \end{aligned}$$

3.2 The Concept of Dimension

We now want to introduce the concept of the *dimension* of a vector space, and to talk about dimensions of subspaces and intersections of subspaces. Basic to this is a rather "technical" lemma, the *basis extension theorem*. Later on, when you are fully conversant with the basic concepts of linear algebra, this will be absorbed into your general store of knowledge, and you will perhaps forget that this item once had a special name — the "basis extension theorem." But at present it is the key to all the remaining concepts and results of this section (not to forget the exercises).

Basis extension theorem: Let V be a vector space over $\mathbb{F}$ and let $v_1, \ldots, v_r$, $w_1, \ldots, w_s$ be vectors of V. If $(v_1, \ldots, v_r)$ is linearly independent and $L(v_1, \ldots, v_r, w_1, \ldots, w_s) = V$, then by suitably chosen vectors from $(w_1, \ldots, w_s)$ one can extend $(v_1, \ldots, v_r)$ to a basis of V.

As a corollary we will obtain the exchange lemma.

Exchange lemma: If $(v_1, \ldots, v_n)$ and $(w_1, \ldots, w_m)$ are bases of V, then for each v_i there exists some w_j, so that on replacing v_i by w_j in $(v_1, \ldots, v_n)$ we still have a basis.

Let us postpone the proofs of both the basis extension theorem and its consequence the exchange lemma until Section 3.4 — not that these proofs are difficult, but because I will then go into certain questions of formulation, which would make dry reading if you have not first seen how useful both results actually are. So let us now turn to applications of the basis extension theorem and the exchange lemma.

Theorem 1: If $(v_1, \ldots, v_n)$ and $(w_1, \ldots, w_m)$ are bases of V, then $m = n$.

PROOF: Assuming that the bases have unequal lengths, say $n > m$, we would be able to apply the exchange lemma repeatedly and trade all vectors of the longer basis against vectors of the shorter, obtaining a long basis made of vectors of the short one, in which therefore at least one vector would appear twice, contradicting the linear independence of the basis. □

Thus any two bases of one and the same vector space are equally long, and Theorem 1 justifies the following definition.

Definition: If the vector space V over $\mathbb{F}$ has a basis $(v_1, \ldots, v_n)$, then n is called the ***dimension*** of V, abbreviated as $\dim V$.

Note: $\dim \mathbb{F}^n = n$, because the canonical basis has length n.

To decide if a given r-tuple $(v_1, \ldots, v_r)$ of vectors in V is linearly dependent or independent usually requires some calculations. It is thus nice to know that in certain cases you need no calculation at all: *each* r-tuple with $r > \dim V$ is linearly dependent.

Theorem 2: Let $v_1, \ldots, v_r$ be vectors in V and $r > \dim V$. Then $(v_1, \ldots, v_r)$ is linearly dependent.

PROOF: Let $(w_1, \ldots, w_n)$ be a basis of V. Then $L(w_1, \ldots, w_n) = V$ and as a consequence $L(v_1, \ldots, v_r, w_1, \ldots, w_n) = V$. Were $(v_1, \ldots, v_r)$ to be linearly *independent* we could use the basis extension theorem to extend $(v_1, \ldots, v_r)$ to a basis (by taking vectors from $(w_1, \ldots, w_n)$). This basis would have length at least equal to r, contradicting $r > \dim V$. □

Thus, if one has to decide on the linear dependence or independence of some r-tuple in V, one will routinely first look if perhaps $r > \dim V$. Four vectors in $\mathbb{R}^3$ are always linearly dependent!

Theorem 2 also enables us to see that there exist vector spaces that have no (finite) basis, and hence for which no dimension is defined. For this we consider an example of a real vector space, already mentioned in Chapter 2. Let M be the real vector space of continuous functions on $[-1, 1]$. For each integer $n > 0$ let $f_n \in M$ be the function with the graph displayed in Fig. 22.

Then for each k the k-tuple $(f_1, \ldots, f_k)$ is linearly independent (because $\lambda_1 f_1 + \cdots + \lambda_k f_k$ takes the value λ_i at the point $\frac{1}{2}(\frac{1}{i} + \frac{1}{i+1})$ for $1 \leq i \leq k$). If now M had a basis $(v_1, \ldots, v_n)$, then (by Theorem 2) $k \leq n$, and this is clearly not possible for all $k > 0$. There is an extended notion of possibly infinite bases and one can show that in this sense *any* vector space has a basis. We do not wish to explore this further here, but we adopt the following terminology.

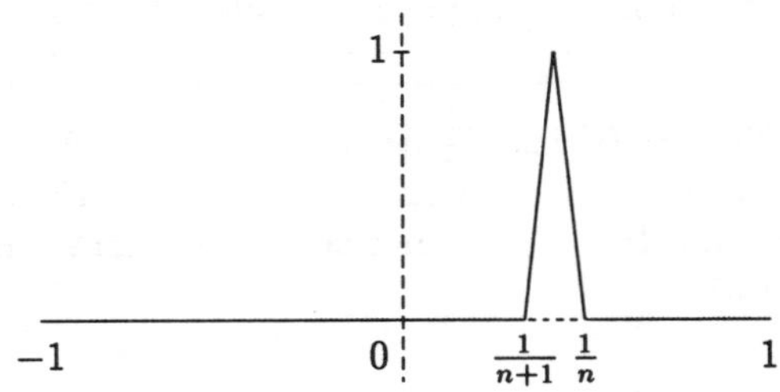

Fig. 22. Graph of the function f_n

Definition: If V possesses no basis $(v_1, \ldots, v_n)$ for $0 \leq n < \infty$, then V is called an ***infinite-dimensional*** vector space, and we write $\dim V = \infty$.

The final object of study in this section is the *dimension of subspaces* of finite-dimensional vector spaces. As answer to the obvious first question we have the following remark.

Remark 3: If V is finite-dimensional and $U \subset V$ is a subspace, then U is also finite-dimensional.

PROOF: If $(v_1, \dots, v_r)$ is linearly independent, then $r \leq \dim V$ by Theorem 2. Hence there exists some *maximal* r, for which one can find a linearly independent r-tuple $(v_1, \dots, v_r)$ in U. But such an r-tuple will then also satisfy $L(v_1, \dots, v_r) = U$ and hence be a basis of U! To see this, note that for each $u \in U$ the $(r+1)$-tuple $(v_1, \dots, v_r, u)$ is linearly dependent, which means that there must be some non-trivial linear combination

$$\lambda_1 v_1 + \cdots + \lambda_r v_r + \lambda u = 0,$$

and we know $\lambda \neq 0$, since otherwise $\lambda_1 v_1 + \cdots + \lambda_r v_r = 0$ would be a nontrivial linear combination. Therefore

$$u = -\frac{\lambda_1}{\lambda} v_1 - \cdots - \frac{\lambda_r}{\lambda} v_r \in L(v_1, \dots, v_r),$$

and we have found a basis for U, showing that it is finite-dimensional. □

A basis $(v_1, \dots, v_r)$ of U can always be extended to a basis of V: just apply the basis extension theorem to $(v_1, \dots, v_r, w_1, \dots, w_n)$, where $(w_1, \dots, w_n)$ is a basis of V. If $U \subsetneqq V$, the basis $(v_1, \dots, v_r)$ must be genuinely lengthened, from which follows the next remark.

Remark 4: If U is a subspace of the finite-dimensional vector space V, then $\dim U < \dim V$ is equivalent to $U \neq V$.

Now let U_1 and U_2 be two subspaces of V. Then $U_1 \cap U_2$ is also a subspace, and we want to try to make some assertion about its dimension. First, one notes that $\dim(U_1 \cap U_2)$ cannot depend on $\dim U_1$ and $\dim U_2$ alone (see Figs. 23a and b).

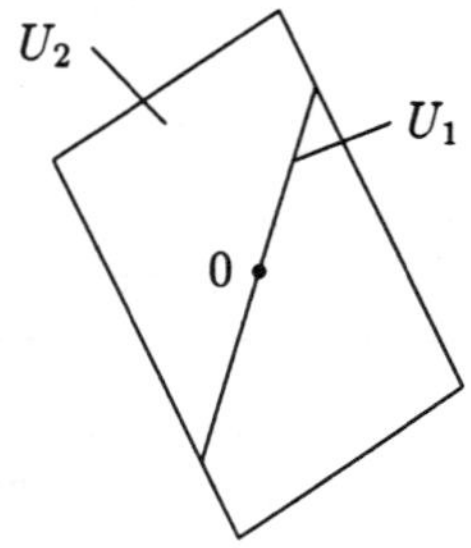

Fig. 23a. $\dim U_1 \cap U_2 = 1$

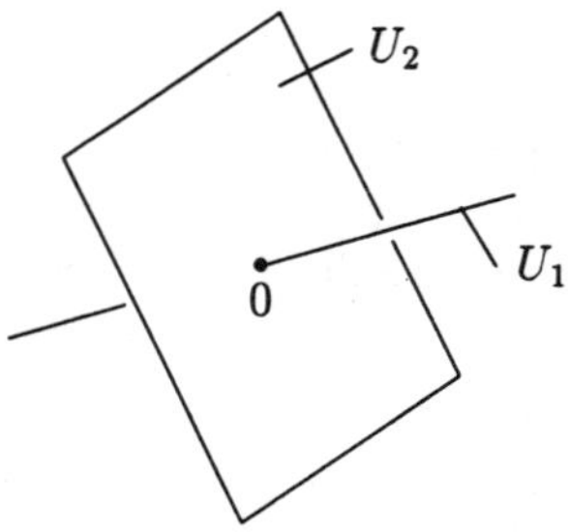

Fig. 23b. $\dim U_1 \cap U_2 = 0$

Thus the relative position of the two subspaces to each other also plays a role. How can one make this precise? To start, we introduce the concept of the *sum of two subspaces*.

Definition: If U_1, U_2 are subspaces of V,

$$U_1 + U_2 := \{x + y \mid x \in U_1,\ y \in U_2\} \subset V$$

is called the ***sum*** of U_1 and U_2.

Of course the sum $U_1 + U_2$ is again a subspace. In order to get used to this new concept, consider why, for example, the statements $U + U = U$, $U + \{0\} = U$, and $U \subset U + U'$ are correct. If you have more time, also try to understand why $U + U' = U \iff U' \subset U$.

Theorem 3 (Dimension formula for subspaces): Let U_1 and U_2 be finite-dimensional subspaces of V, then

$$\dim(U_1 \cap U_2) + \dim(U_1 + U_2) = \dim U_1 + \dim U_2.$$

PROOF: This proceeds with the help of the basis extension theorem. First choose a basis $(v_1, \ldots, v_r)$ of $U_1 \cap U_2$ and then extend it once to a basis $(v_1, \ldots, v_r, w_1, \ldots, w_s)$ of U_1, and a second time to a basis $(v_1, \ldots, v_r, z_1, \ldots, z_t)$ of U_2.

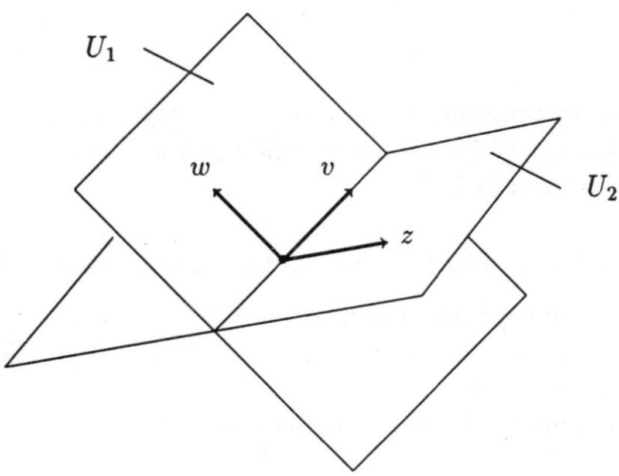

Fig. 24. Summary of the proof of the dimension formula: first choose v, then w, and then z. Now count.

Then $(v_1, \ldots, v_r, w_1, \ldots, w_s, z_1, \ldots, z_t)$ is a basis of $U_1 + U_2$. Why? Clearly $L(v_1, \ldots, z_t) = U_1 + U_2$, and we have to show only that $(v_1, \ldots, z_t)$ is linearly independent. Suppose therefore that

$$\lambda_1 v_1 + \cdots + \lambda_r v_r + \mu_1 w_1 + \cdots + \mu_s w_s + \nu_1 z_1 + \cdots + \nu_t z_t = 0.$$

Then $\nu_1 z_1 + \cdots + \nu_t z_t \in U_1 \cap U_2$, since it is certainly in U_2 and lies in U_1 as a consequence of the equation $\nu_1 z_1 + \cdots + \nu_t z_t = -\lambda_1 v_1 - \cdots - \mu_s w_s$. But then for suitable $\alpha_1, \ldots, \alpha_r$ we have $\nu_1 z_1 + \cdots + \nu_t z_t = \alpha_1 v_1 + \cdots + \alpha_r v_r$ because $(v_1, \ldots, v_r)$ is a basis of $U_1 \cap U_2$. From this it follows that all the ν's and α's are zero, because $(v_1, \ldots, v_r, z_1, \ldots, z_t)$ is linearly independent. Hence

$$\lambda_1 v_1 + \cdots + \lambda_r v_r + \mu_1 w_1 + \cdots + \mu_s w_s = 0,$$

implying that the λ's and μ's also vanish. These two steps show the desired linear independence of $(v_1, \ldots, z_t)$. Counting dimensions, we have $\dim U_1 \cap U_2 = r$, $\dim U_1 = r+s$, $\dim U_2 = r+t$, $\dim U_1 + U_2 = r+s+t$, from which the formula to be proved follows, namely $\dim U_1 \cap U_2 + \dim U_1 + U_2 = \dim U_1 + \dim U_2$. □

3.3 Test

(1) For which of the following objects does the description "linearly dependent" or "linearly independent" make sense?

☐ An n-tuple $(v_1, \ldots, v_n)$ of elements of a vector space
☐ An n-tuple $(v_1, \ldots, v_n)$ of real vector spaces
☐ A linear combination $\lambda_1 v_1 + \cdots + \lambda_n v_n$

(2) Let $v_1, \ldots, v_n \in V$. What does $L(v_1, \ldots, v_n) = V$ mean?

☐ Each linear combination $\lambda_1 v_1 + \cdots + \lambda_n v_n$ is an element of V.
☐ Each element of V is a linear combination $\lambda_1 v_1 + \cdots + \lambda_n v_n$.
☐ The dimension of V is n.

(3) For linearly independent triples (v_1, v_2, v_3) of vectors in V,

☐ (v_1, v_2) is always linearly dependent.
☐ (v_1, v_2) may or may not be linearly dependent, depending on the choice of (v_1, v_2, v_3).
☐ (v_1, v_2) is always linearly independent.

(4) The ith vector of the canonical basis of $\mathbb{F}^n$ is defined by

☐ $e_i = (0, \ldots, i, \ldots, 0)$.
☐ $e_i = (0, \ldots, 1, \ldots, 0)$.
☐ $e_i = (1, \ldots, 1, \ldots, 0)$.

(5) Which of the following statements implies the linear independence of the n-tuple $(v_1, \ldots, v_n)$ of elements of V?

☐ $\lambda_1 v_1 + \cdots + \lambda_n v_n = 0$ only if $\lambda_1 = \lambda_2 = \cdots = \lambda_n = 0$.
☐ If $\lambda_1 = \cdots = \lambda_n = 0$, then $\lambda_1 v_1 + \cdots + \lambda_n v_n = 0$.
☐ $\lambda_1 v_1 + \cdots + \lambda_n v_n = 0$ for all $(\lambda_1, \ldots, \lambda_n) \in \mathbb{F}^n$.

(6) In the basis extension theorem a linearly independent r-tuple of vectors is extended to a basis by means of vectors from an s-tuple of vectors (assuming that the vectors taken together generate the space). What does the basis extension theorem say in the case $r = 0$?

□ If $(w_1, \ldots, w_s)$ is an s-tuple of vectors in V and $L(w_1, \ldots, w_s) = V$, then one can extend $w_1, \ldots, w_s$ to a basis.
□ If $(w_1, \ldots, w_s)$ is linearly independent, then there exists a basis consisting of vectors from $(w_1, \ldots, w_s)$.
□ If $L(w_1, \ldots, w_s) = V$, there exists a basis consisting of vectors from $(w_1, \ldots, w_s)$.

(7) The vector space $V = \{0\}$ consisting of zero

□ has the basis (0) □ has the basis $\varnothing$ □ has no basis

(8) If one were to define $U_1 - U_2 := \{x - y \mid x \in U_1, y \in U_2\}$ for subspaces U_1, U_2 of V, then one would have

□ $U_1 - U_1 = \{0\}$.
□ $(U_1 - U_2) + U_2 = U_1$.
□ $U_1 - U_2 = U_1 + U_2$.

(9) One always has

□ $(U_1 + U_2) + U_3 = U_1 + (U_2 + U_3)$.
□ $U_1 \cap (U_2 + U_3) = (U_1 \cap U_2) + (U_1 \cap U_3)$.
□ $U_1 + (U_2 \cap U_3) = (U_1 + U_2) \cap (U_1 + U_3)$.

(10) Subspaces U_1, U_2 of V are said to be transverse (to each other) if $U_1 + U_2 = V$. One calls $\operatorname{codim} U := \dim V - \dim U$ the codimension of U in V. For transverse U_1, U_2, one has

□ $\dim U_1 + \dim U_2 = \dim U_1 \cap U_2$.
□ $\dim U_1 + \dim U_2 = \operatorname{codim} U_1 \cap U_2$.
□ $\operatorname{codim} U_1 + \operatorname{codim} U_2 = \operatorname{codim} U_1 \cap U_2$.

3.4 Proof of the Basis Extension Theorem and the Exchange Lemma

A section for mathematicians

Proof of the Basis Extension Theorem: The theorem reads: "If V is a vector space over $\mathbb{F}$, if $L(v_1, \ldots, v_r, w_1, \ldots, w_s) = V$, and if $(v_1, \ldots, v_r)$ is linearly independent, then by suitably chosen vectors from $(w_1, \ldots, w_s)$ one can extend $(v_1, \ldots, v_r)$ to a basis of V." Here we allow the cases $r = 0$ and $s = 0$. (The empty "0-tuple" counts as being linearly independent and $L(\varnothing) = \{0\}$.)

We carry out the proof by induction on s. In the case $s = 0$ (start of the induction) there is nothing to prove, because $(v_1, \ldots, v_r)$ is already a basis. We must show that if the theorem holds for $s = n$ (inductive assumption) then it also holds for $s = n+1$. Suppose therefore that we have $(v_1, \ldots, v_r, w_1, \ldots, w_{n+1})$ with $L(v_1, \ldots, w_{n+1}) = V$ and $(v_1, \ldots, v_r)$ linearly independent.

If $L(v_1, \ldots, v_r) = V$ already, then $(v_1, \ldots, v_r)$ is a basis and the assertion is proved in this case. Suppose therefore that $L(v_1, \ldots, v_r) \neq V$. Then at least one of the w_i is not contained in $L(v_1, \ldots, v_r)$, because otherwise $L(v_1, \ldots, v_r)$ would be $L(v_1, \ldots, v_r, w_1, \ldots, w_{n+1})$, which is V.

For such a w_i the $(r+1)$-tuple $(v_1, \ldots, v_r, w_i)$ is then linearly independent since from $\lambda_1 v_1 + \cdots + \lambda_r v_r + \lambda w_i = 0$ it follows first that $\lambda = 0$ (because $w_i \notin L(v_1, \ldots, v_r)$) and then that $\lambda_1 = \cdots = \lambda_r = 0$ (because $(v_1, \ldots, v_r)$ is linearly independent). By the inductive assumption one can now extend $(v_1, \ldots, v_r, w_i)$ to a basis of V by choice of suitable vectors from the n-tuple $(w_1, \ldots, w_{i-1}, w_{i+1}, \ldots, w_{n+1})$, so that the desired extension of $(v_1, \ldots, v_r)$ to a basis of V has been achieved. □

With this proof on hand, let us discuss some questions of mathematical formulation. In the theorem one has "... *then by suitably chosen vectors from* $(w_1, \ldots, w_s)$ *one can extend* $(v_1, \ldots, v_r)$ *to a basis of* V." In more complicated mathematical situations one cannot manage with such a verbal description (which, however, does very well in simple cases!). What would more formal notation look like? If one extends $(v_1, \ldots, v_r)$ by vectors $a_1, \ldots, a_k$, one naturally obtains the $(r+k)$-tuple $(v_1, \ldots, v_r, a_1, \ldots, a_k)$. But how does one write that $a_1, \ldots, a_k$ are taken from $(w_1, \ldots, w_s)$? One cannot simply write $w_1, \ldots, w_k$, since this specifically describes the *first* k vectors of the s-tuple $(w_1, \ldots, w_s)$.

> If one wants to describe a k-tuple of vectors chosen from $(w_1, \ldots, w_s)$, one must use *indexed indices.* Each such k-tuple can be written as $(w_{i_1}, \ldots, w_{i_k})$, where the i_α are integers with $1 \leq i_\alpha \leq s$. If in addition one requires that no w_i is used more than once, one must suppose that the i_α are pairwise distinct, that is, $i_\alpha \neq i_\beta$ for $\alpha \neq \beta$.

We could then formulate the basis extension theorem as follows: "If V is a vector space over $\mathbb{F}$, $L(v_1, \ldots, v_r, w_1, \ldots, w_s) = V$ and $(v_1, \ldots, v_r)$ is linearly independent, then either $(v_1, \ldots, v_r)$ is a basis or there exist pairwise distinct integers $i_1, \ldots, i_k$, with $1 \leq i_\alpha \leq s$ for $\alpha = 1, \ldots, k$, such that $(v_1, \ldots, v_r, w_{i_1}, \ldots, w_{i_k})$ is a basis."

A second point of the proof that we wish formally to "fill out" concerns the assertion that from $w_1, \ldots, w_{n+1} \in L(v_1, \ldots, v_r)$ it follows that $L(v_1, \ldots, v_r) = L(v_1, \ldots, v_r, w_1, \ldots, w_{n+1})$. Indeed this is clear, since on one hand each $\lambda_1 v_1 + \cdots + \lambda_r v_r \in L(v_1, \ldots, v_r)$ can be written as the linear combination $\lambda_1 v_1 + \cdots + \lambda_r v_r + 0w_1 + \cdots + 0w_{n+1}$, and conversely if $v = \lambda_1 v_1 + \cdots + \lambda_r v_r + \mu_1 w_1 + \cdots + \mu_{n+1} w_{n+1} \in L(v_1, \ldots, w_{n+1})$ is given, we need "only" to express the w_i as linear combinations of the v_i to see that v is also a linear combination of the v_i. But if we really want to write everything out we must use *double indices.*

If each w_i must be written as a linear combination of $(v_1, \ldots, v_r)$, the coefficients must be so labeled that one immediately sees to which w_i and also to which v_j they refer. One can, for example, write the coefficient as λ^i_j, $i = 1, \ldots, n{+}1$, $j = 1, \ldots, r$, where i is an *upper index* and not an ith power. But one can also write the two indices next to each other: λ_{ij}.

With this notation we can then say: "If $w_1, \ldots, w_{n+1} \in L(v_1, \ldots, v_r)$, then $w_i = \lambda_{i1} v_1 + \cdots + \lambda_{ir} v_r$, $i = 1, \ldots, n{+}1$, for suitable $\lambda_{ij} \in \mathbb{F}$."

For a linear combination v of $(v_1, \ldots, v_r, w_1, \ldots, w_{n+1})$ we then have

$$\begin{aligned} v &= \lambda_1 v_1 + \cdots + \lambda_r v_r + \mu_1 w_1 + \cdots + \mu_{n+1} w_{n+1} \\ &= \lambda_1 v_1 + \cdots + \lambda_r v_r + \mu_1(\lambda_{11} v_1 + \cdots + \lambda_{1r} v_r) + \cdots + \mu_{n+1}(\lambda_{n+1,1} v_1 + \cdots + \lambda_{n+1,r} v_r) \end{aligned}$$

and collecting terms we obtain

$$v = (\lambda_1 + \mu_1 \lambda_{11} + \cdots + \mu_{n+1}\lambda_{n+1,1}) v_1 + \cdots + (\lambda_r + \mu_1 \lambda_{1r} + \cdots + \mu_{n+1}\lambda_{n+1,r}) v_r$$

and so $v \in L(v_1, \ldots, v_r)$. Please observe that we could *not* have written $w_i = \lambda_{i_1} v_1 + \cdots + \lambda_{i_r} v_r$.

The conclusion of the proof reads: "By the inductive assumption there therefore exist, for some suitable k, pairwise distinct integers i_α, $\alpha = 1, \ldots, k{-}1$, with $1 \leq i_\alpha \leq n+1$ and $i_\alpha \neq i$, such that $(v_1, \ldots, v_r, w_i, w_{i_1}, \ldots, w_{i_{k-1}})$ is a basis of V. This achieves the desired extension of $(v_1, \ldots, v_r)$ to a basis of V."

Proof of the Exchange Lemma: Let $(v_1, \ldots, v_n)$ and $(w_1, \ldots, w_m)$ be two bases of V and $i \in \{1, \ldots, n\}$ chosen to be fixed. Then there must be some j so that $w_j \notin L(v_1, \ldots, v_{1-1}, v_{i+1}, \ldots, v_n)$. Otherwise we would have $L(v_1, \ldots, v_{i-1}, v_{i+1}, \ldots, v_n) \supset L(w_1, \ldots, w_m) = V$, which cannot be the case since the linear independence of $(v_1, \ldots, v_n)$ implies that v_i cannot be a linear combination of $(v_1, \ldots, v_{i-1}, v_{i+1}, \ldots, v_n)$.

For such a j the n-tuple $(v_1, \ldots, v_{i-1}, w_j, v_{i+1}, \ldots, v_n)$ is then linearly independent, since from a relation

$$\lambda_1 v_1 + \cdots + \lambda_{i-1} v_{i-1} + \mu w_j + \lambda_{i+1} v_{i+1} + \cdots + \lambda_n v_n = 0,$$

it follows first that $\mu = 0$ and then that $\lambda_1 = \cdots = \lambda_n = 0$.

Since $L(v_1, \ldots, v_{i-1}, w_j, v_{i+1}, \ldots, v_n, v_i) = V$, we can apply the basis extension theorem to see that either $(v_1, \ldots, v_{i-1}, w_j, v_{i+1}, \ldots, v_n)$ is itself already a basis, or it becomes a basis on extending it by v_i. But the latter cannot occur, since w_j, just as any $v \in V$, is a linear combination of $(v_1, \ldots, v_n)$, and hence $(v_1, \ldots, v_{i-1}, w_j, v_{i+1}, \ldots, v_n, v_i)$ is linearly dependent. Therefore $(v_1, \ldots, v_{i-1}, w_j, v_{i+1}, \ldots, v_n)$ is a basis. □

3.5 The Vector Product

A section for physicists

In mathematics and physics several sorts of "products" of two or more vectors are considered. We have already heard of *inner products*, but there also exist a *cross* or *vector product*, a *tensor product*, an *exterior* or *alternating product*, a *Lie bracket product* and others. On closer inspection these turn out to be very different. What additional structures and assumptions one needs, whether a product is again a vector or a scalar, whether it must lie in the same vector space as the factors or not, whether the factors have indeed to come from the same vector space, whether the product remains unchanged on switching the factors, or changes sign, or changes in some more drastic fashion, whether in the presence of several factors one can insert brackets in an arbitrary way (as one can do with numbers) — all this varies from case to case.

There is one thing, however, of which one can be rather certain about "products": they are *multilinear*, that is, if I replace one factor v by a sum $v_1 + v_2$, without altering the others, I obtain the sum of the products obtained by using v_1 and v_2 together with the remaining factors. In the same way, if I replace v by λv, $\lambda \in \mathbb{R}$, again without altering the other factors, I obtain the original product multiplied by λ. Therefore, if one writes the factors as linear combinations of basis vectors, one can work out their product knowing only the products of the basis vectors.

In this section we wish to consider the vector product for physical vectors (see Section 2.6), and in doing this we again start with the "dimensionless" physical vector space $\mathcal{U}$, understood according to choice as $\mathcal{U}_O$ or $\mathcal{U}_{\text{free}}$. If $\vec{u}$ and $\vec{v}$ are two vectors from $\mathcal{U}$, their vector product $\vec{u} \times \vec{v}$ is also in $\mathcal{U}$; the vector product is a bilinear map $\mathcal{U} \times \mathcal{U} \to \mathcal{U}$, $(\vec{u}, \vec{v}) \mapsto \vec{u} \times \vec{v}$. In order to define and calculate the vector product, we must know which of the orthonormal bases $\hat{x}, \hat{y}, \hat{z}$ of $\mathcal{U}$ are "right-" and which are "left-handed." If $\hat{x}, \hat{y}, \hat{z}$, in this order, point in the direction of the thumb, index finger, and middle finger of the right hand, $\hat{x}, \hat{y}, \hat{z}$ is right-handed. Put another way: if the right thumb points in the direction $\hat{z}$ and the rotational direction indicated by the fingers is the one that moves $\hat{x}$ through a quarter turn into $\hat{y}$, then $(\hat{x}, \hat{y}, \hat{z})$ is right-handed. Refer to Figs. 25a and b for illustrations of these concepts.

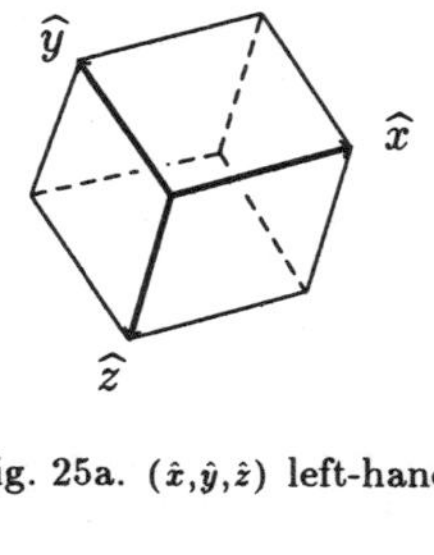

Fig. 25a. $(\hat{x},\hat{y},\hat{z})$ left-handed

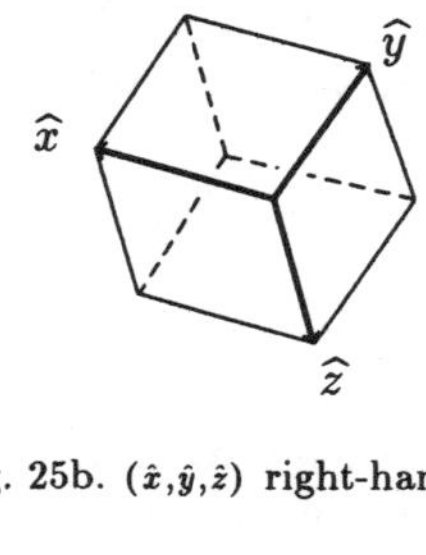

Fig. 25b. $(\hat{x},\hat{y},\hat{z})$ right-handed

One could easily give equivalent descriptions in turns of a screw, steering wheel, taps, celestial directions, etc. The definition is neither unclear nor unscientific, but is nonmathematical in that it depends essentially on actual physical space, since I cannot insert my hand into $\mathbb{R}^3$ or indeed into any other abstract three-dimensional Euclidean vector space $(V, \langle\, , \rangle)$. In order to define the vector product $V \times V \to V$, one must first mathematically imitate the natural phenomenon of right-handedness by means of an additional structure in V called an "orientation." At present we will not go into this; those already acquainted with the concept of orientation can read the following theorem in terms of an arbitrary three-dimensional oriented Euclidean vector space $\mathcal{U}$.

Theorem (and definition of the vector product): There exists a unique bilinear composition $\mathcal{U} \times \mathcal{U} \to \mathcal{U}$, written as $(\vec{u}, \vec{v}) \mapsto \vec{u} \times \vec{v}$, with the following two properties:

(1) The composition is skew-symmetric, i.e., one has $\vec{u} \times \vec{v} = -\vec{v} \times \vec{u}$.

(2) If $\vec{u}$ and $\vec{v}$ are mutually perpendicular unit vectors, then $\vec{u} \times \vec{v}$ extends the pair $(\vec{u}, \vec{v})$ to a right-handed orthonormal basis $(\vec{u}, \vec{v}, \vec{u}\times\vec{v})$.

This composition is called the ***vector product.***

The proof will help us to understand the vector product from both the computational and geometric points of view. We begin with the remark that, because of the skew-symmetry, we must always have the following condition.

(3) $$\vec{u} \times \vec{u} = 0$$

Also, because of (2), for a right-handed orthonormal basis $(\hat{x}, \hat{y}, \hat{z})$, the following formulae must hold for all compositions satisfying (1) and (2).

(4) $$\begin{aligned} \hat{x} \times \hat{y} &= -\hat{y} \times \hat{x} = \hat{z}, \\ \hat{y} \times \hat{z} &= -\hat{z} \times \hat{y} = \hat{x}, \\ \hat{z} \times \hat{x} &= -\hat{x} \times \hat{z} = \hat{y} \end{aligned}$$

But then we know the product for the basis vectors and, because of bilinearity, for all vectors. Indeed, it is now clear that we can obtain our composition only by means of the following formula.

(5) $$\begin{aligned} \vec{u} \times \vec{v} &= (u_x\hat{x} + u_y\hat{y} + u_z\hat{z}) \times (v_x\hat{x} + v_y\hat{y} + v_z\hat{z}) \\ &:= (u_yv_z - u_zv_y)\hat{x} + (u_zv_x - u_xv_z)\hat{y} + (u_xv_y - u_yv_x)\hat{z} \end{aligned}$$

The product *defined* by (5) is clearly bilinear, and it satisfies (1). We still need to check (2). By the way, readers who can already work out three-row determinants, either because they have browsed ahead in Chapter 6 or because they are not really beginners and only want to polish up their vector product, will recognize that one can also formally write (5) as follows.

(5′) $$\vec{u} \times \vec{v} = \det \begin{pmatrix} u_x & u_y & u_z \\ v_x & v_y & v_z \\ \hat{x} & \hat{y} & \hat{z} \end{pmatrix}$$

Such readers will also easily verify the following formula.

(5″) $$(\vec{u} \times \vec{v}) \cdot \vec{w} = \det \begin{pmatrix} u_x & u_y & u_z \\ v_x & v_y & v_z \\ w_x & w_y & w_z \end{pmatrix}$$

The latter is very useful, and full of geometric significance, since this mixed product of the three factors $\vec{u}, \vec{v}$, and $\vec{w}$ determines the *volume* of the parallelotope (three-dimensional generalization of a parallelogram) spanned by $\vec{u}, \vec{v}$, and $\vec{w}$ — up to a sign, which is given by the "handedness" of the three vectors.

But we should not pretend to know determinants already, and so instead let us use (5) to give an elementary derivation of the next formula.

(6) $$(\vec{u} \times \vec{v}) \cdot (\vec{u'} \times \vec{v'}) = (\vec{u} \cdot \vec{u'})(\vec{v} \cdot \vec{v'}) - (\vec{u} \cdot \vec{v'})(\vec{v} \cdot \vec{u'})$$

One can either work it out, or argue as follows: both sides in the desired equation (6) are linear in each of the four variables, and so we only need to check (6) for $\vec{u}, \vec{u'}, \vec{v}, \vec{v'} \in \{\hat{x}, \hat{y}, \hat{z}\}$. For $\vec{u} = \vec{v}$ or $\vec{u'} = \vec{v'}$, both sides are zero anyway, so we may suppose that $\vec{u} \neq \vec{v}$ and $\vec{u'} \neq \vec{v'}$. Therefore, without loss of generality, and because of symmetry, it remains only to check (6) for $(\vec{u}, \vec{v}, \vec{u'}, \vec{v'})$ equal to $(\hat{x}, \hat{y}, \hat{x}, \hat{y})$ or $(\hat{x}, \hat{y}, \hat{y}, \hat{z})$. In the first case, both sides equal 1, and in the second case, both sides are zero, verifying (6).

Readers with a knowledge of determinants will prefer to read (6) as shown below.

(6′) $$(\vec{u} \times \vec{v}) \cdot (\vec{u'} \times \vec{v'}) = \det \begin{pmatrix} \vec{u} \cdot \vec{u'} & \vec{u} \cdot \vec{v'} \\ \vec{v} \cdot \vec{u'} & \vec{v} \cdot \vec{v'} \end{pmatrix}$$

Similar useful formulae, which also follow from (5), appear here as (7)–(9).

(7) $$(\vec{u} \times \vec{v}) \cdot \vec{w} = (\vec{v} \times \vec{w}) \cdot \vec{u} = (\vec{w} \times \vec{u}) \cdot \vec{v}$$

This implies (8) because of (3).

(8) $$\vec{u} \perp (\vec{u} \times \vec{v}) \quad \text{and} \quad \vec{v} \perp (\vec{u} \times \vec{v})$$

Furthermore, consider (9).

$$(9) \qquad \vec{u} \times (\vec{v} \times \vec{w}) = (\vec{u} \cdot \vec{w})\vec{v} - (\vec{u} \cdot \vec{v})\vec{w}.$$

As with formula (6), one proves (7) and (9) either by computation or by relying on the linearity in the three factors. Because of linearity, one may assume without loss of generality that $\vec{u}, \vec{v}, \vec{w} \in \{\hat{x}, \hat{y}, \hat{z}\}$. For $\vec{u} = \vec{v} = \vec{w}$ both formulae are trivially true; hence on grounds of symmetry only the cases $(\hat{x}, \hat{x}, \hat{y})$ and $(\hat{x}, \hat{y}, \hat{z})$ for $(\vec{u}, \vec{v}, \vec{w})$ remain to be verified, and so forth. If one uses determinants, (7) is immediately clear from (5″) anyway.

In order to complete the proof of our theorem we have still to show that (2) also follows from (5). So now suppose that $\vec{u}$ and $\vec{v}$ are two mutually perpendicular unit vectors. We have $|\vec{u} \times \vec{v}| = 1$ by (6) and $\vec{u} \times \vec{v}$ perpendicular to $\vec{u}$ and $\vec{v}$ by (8). But why is $(\vec{u}, \vec{v}, \vec{u} \times \vec{v})$ right-handed?

For readers who are used to the concept of orientation and who in $\mathcal{U}$ see only an oriented three-dimensional Euclidean vector space, this follows from (5″) applied to $\vec{w} = \vec{u} \times \vec{v}$. (The determinant is then positive, which implies the right-handedness of $(\vec{u}, \vec{v}, \vec{u} \times \vec{v})$ in the mathematical sense, since $(\hat{x}, \hat{y}, \hat{z})$ was assumed to be right-handed.)

With physically defined right-handedness we argue as follows. Let $\vec{w}$ be the unit vector that extends $(\vec{u}, \vec{v})$ to a right-handed orthonormal system, thus $\vec{w} = \pm \vec{u} \times \vec{v}$, but as yet we do not know the sign. We can however change $(\hat{x}, \hat{y}, \hat{z})$ into $(\vec{u}, \vec{v}, \vec{w})$ by a continuous motion ("of the right hand"). Let $(\vec{u}(t), \vec{v}(t), \vec{w}(t))$ denote the moving system at time t. Then we certainly have $\vec{u}(t) \times \vec{v}(t) = \pm \vec{w}(t)$, so that $|\vec{u}(t) \times \vec{v}(t) - \vec{w}(t)|$ is either 0 or 2. But initially this quantity is zero because $\hat{x} \times \hat{y} = \hat{z}$, and thus by continuity it also takes the value zero at the end, implying that $\vec{u} \times \vec{v} = \vec{w}$. □

With this the theorem is proved and the vector product $\mathcal{U} \times \mathcal{U} \to \mathcal{U}$ is defined. We have also learned that it is bilinear and has properties (1) through (9).

Along with the vector product in $\mathcal{U}$, the vector product of arbitrary physical vectors is defined in a canonical manner by

$$\begin{aligned} \mathcal{U}[a] \times \mathcal{U}[b] &\longrightarrow \mathcal{U}[\mathrm{ab}] \\ (a\vec{u}, b\vec{v}) &\longmapsto \mathrm{ab}(\vec{u} \times \vec{v}). \end{aligned}$$

Its properties follow immediately from those of the vector product in the oriented three-dimensional Euclidean vector space $\mathcal{U}$, which once again provides the link between abstract linear algebra and physical vector calculus.

To define the vector product in an initially *nonoriented* abstract three-dimensional Euclidean vector space, one must first "orient" the space. For this one must arbitrarily choose some basis to be right-handed or positively oriented — which other ones then count as right-handed is intuitively clear and can mathematically be expressed by determinantal or motion conditions. In the vector space of number triples it is customary to label the canonical basis (e_1, e_2, e_3) as positively oriented; by (5) the vector product in $\mathbb{R}^3$ is then given by (10).

$$(10) \qquad \begin{pmatrix} x_1 \\ x_2 \\ x_3 \end{pmatrix} \times \begin{pmatrix} y_1 \\ y_2 \\ y_3 \end{pmatrix} = \begin{pmatrix} x_2 y_3 - x_3 y_2 \\ x_3 y_1 - x_1 y_3 \\ x_1 y_2 - x_2 y_1 \end{pmatrix} \in \mathbb{R}^3$$

Correspondingly (1) through (9) hold.

Finally, we wish to derive the usual geometric description of the vector product from (6) and (7). With $\vec{u} = \vec{u'} \neq 0$ and $\vec{v} = \vec{v'} \neq 0$, the next equation follows from (6).

$$(11) \qquad \begin{aligned} |\vec{u} \times \vec{v}| &= \sqrt{|\vec{u}|^2 |\vec{v}|^2 - (\vec{u} \cdot \vec{v})^2} \\ &= |\vec{u}||\vec{v}|\sqrt{1 - \cos^2 \alpha(\vec{u}, \vec{v})} \\ &= |\vec{u}||\vec{v}| \sin \alpha(\vec{u}, \vec{v}) \end{aligned}$$

Thus, for position vectors $\vec{u}, \vec{v} \in \mathcal{A}_O$, the magnitude

$$|\vec{u} \times \vec{v}| \in \mathbb{R}[\mathrm{cm}^2]$$

is just the area of the parallelogram spanned by $\vec{u}$ and $\vec{v}$ (see Fig. 26). This can be extended to other cases: one calls $|\vec{u}||\vec{v}| \sin \alpha(\vec{u}, \vec{v})$ the "area" of the parallelogram, even when $\vec{u}, \vec{v}$ are not position vectors and $|\vec{u}||\vec{v}|$ lies in some other scalar domain, or in mathematical linear algebra in $\mathbb{R}$. By (8) $\vec{u} \times \vec{v}$ is always perpendicular to $\vec{u}$ and $\vec{v}$, and if $\vec{u} \times \vec{v} \neq 0$ then $(\vec{u}, \vec{v}, \vec{u} \times \vec{v})$ is right-handed if this concept is generalized in an obvious way from orthonormal to arbitrary bases. One can then make the following statement.

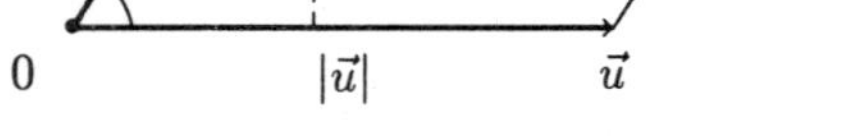

Fig. 26. The area of the parallelogram can be calculated from $|\vec{u}|$, $|\vec{v}|$, and α.

If $\vec{u}$ and $\vec{v}$ are linearly independent, $\vec{u} \times \vec{v}$ is that vector perpendicular to both $\vec{u}$ and $\vec{v}$ with magnitude equal to the area of the parallelogram, which extends $(\vec{u}, \vec{v})$ in a right-handed way. For linearly dependent vectors $\vec{u}, \vec{v}$, the surface area and hence the vector product are of course zero.

3.6 The "Steinitz Exchange Theorem"

Historical aside

In linear algebra textbooks the following theorem is often referred to as the "exchange theorem of Steinitz."

Theorem: *If a vector space V has a basis of p vectors and if $(v_1, \dots, v_r)$ are linearly independent in V, then there exists a basis of p vectors for V, in which $v_1, \dots, v_r$ all occur.*

We have of course implicitly proved this theorem in our present chapter: that there exists some basis containing $v_1, \dots, v_r$ follows from the basis extension theorem, and that this basis has length p follows from Theorem 1. For Steinitz this theorem occurs in a work from 1913, and there reads as follows:

If the module M has a basis of p numbers, and contains r linearly independent numbers $\beta_1, \dots \beta_r$, then it also has a basis of p numbers, among which the numbers $\beta_1, \dots, \beta_r$ all occur.

Translating Steinitz's terminology into ours we obtain the theorem stated above.

A witticism frequently quoted among mathematicians states that, if a theorem is named after someone, then this is a sign that the person so honored was not the first to have proved this theorem. This appears to be true in this case: in Schwerdtfeger [5] I found a footnote on page 23: "This theorem (Austauschsatz) is usually ascribed to E. Steinitz alone. It has been pointed out however by H. G. Forder in his book "The Calculus of Extensions," Cambridge 1941, p. 219, that H. Grassmann had published this theorem in 1862, i.e. 52 years before Steinitz."

Well, Ernst Steinitz, who lived from 1871 to 1928 and was an important algebraist, certainly had no intention to claim any originality for this theorem. The work (in German) in which the theorem occurs is called "Conditionally Convergent Series and Convex Systems," and appeared in the *Journal fuer die reine und angewandte Mathematik* (the so-called *Crelle Journal*), Volume 143 (1913), with a second part following in Volume 144. At the beginning of this work, before attacking the actual subject, Steinitz gives a short introduction to the basic concepts of linear algebra, including the exchange theorem. He even apologizes for this, writing, "The foundations of n-dimensional geometry, which are used here repeatedly, might have been assumed to be known. I have preferred to repeat their derivation. Of course this is just a matter of presentation. I believe that the way chosen here has its advantages and thus may not appear to be superfluous."

So we will certainly not do Steinitz justice if we only remember him as the one who proved the exchange theorem. It is also obvious that such a simple observation as the exchange theorem could no longer be regarded as a noteworthy scientific result in 1913. You have only to think, for example, that the theory of relativity had been conceived in 1905.

You will come across the names of many mathematicians, because their names are attached to concepts and theorems. Do not draw too many conclusions about these mathematicians and the state of knowledge in their time. Sometimes, the theorem is beneath the level of the name (as here for the Steinitz exchange theorem), sometimes, on the other hand, a deep theorem in modern mathematics is named after an old mathematician, who perhaps only proved a special case. And this is what I really wanted to tell you in this "historical aside."

3.7 Exercises

Exercises for mathematicians

3.1: Let V be a real vector space and $a, b, c, d \in V$. Suppose that

$$\begin{aligned} v_1 &= a + b + c + d \\ v_2 &= 2a + 2b + c - d \\ v_3 &= a + b + 3c - d \\ v_4 &= a - c + d \\ v_5 &= -b + c - d \end{aligned}$$

Show that $(v_1, \ldots, v_5)$ is linearly dependent.

One can solve this exercise by expressing one of the v_i as a linear combination of the other four. But there is also a proof in which one does not need to do any calculations.

3.2: Let V be a vector space over $\mathbb{F}$ and U_1, U_2 be subspaces of V. We say that U_1 and U_2 are ***complementary subspaces*** if $U_1 + U_2 = V$ and $U_1 \cap U_2 = \{0\}$.

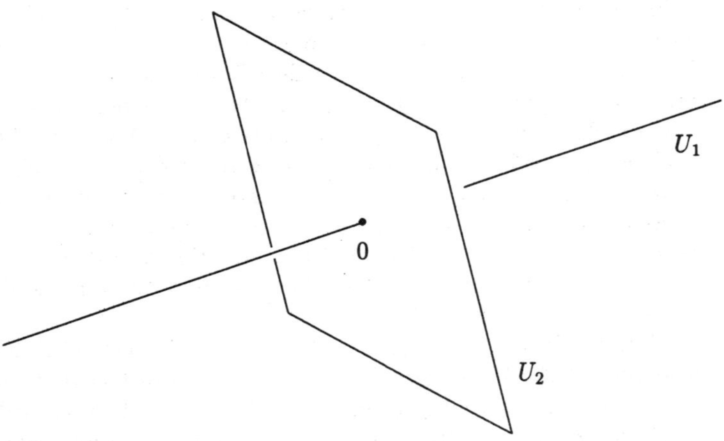

Fig. 27. Example of complementary subspaces in $\mathbb{R}^3$

Of course, $U_1 = V$ and $U_2 = \{0\}$ are also complementary to each other.

Show that if V is an n-dimensional vector space over $\mathbb{F}$ and U_1 is a p-dimensional subspace of V, then there exists a subspace U_2 complementary to U_1, and each such subspace U_2 has dimension $n-p$.

3.3: In Theorem 2 we showed that in a finite-dimensional vector space V a linearly independent r-tuple $(v_1, \dots, v_r)$ can have length at most equal to $\dim V$. Now show that in an infinite dimensional space V there exists some infinite sequence $v_1, v_2, \dots$ of vectors, such that for each r the r-tuple $(v_1, \dots, v_r)$ is linearly independent.

The $*$-exercise

3*: Given a complex vector space V one can make a *real* vector space from it by simply restricting the scalar multiplication $\mathbb{C} \times V \to V$ to $\mathbb{R} \times V$. Since on restriction the concepts "linear hull" and "dimension" take on a new meaning, instead of L, dim, we want to write $L_{\mathbb{C}}$, $\dim_{\mathbb{C}}$, or $L_{\mathbb{R}}$, $\dim_{\mathbb{R}}$, depending on whether V is being considered as a complex or real vector space. *Exercise:* For each $n \geq 0$ determine for which pairs (r, s) of numbers there exists a complex vector space and vectors $(v_1, \dots, v_n)$ in it, such that $r = \dim_{\mathbb{R}} L_{\mathbb{C}}(v_1, \dots, v_n)$ and $s = \dim_{\mathbb{R}} L_{\mathbb{R}}(v_1, \dots, v_n)$.

Exercises for physicists

3.1P: Exercise 3.1 (for mathematicians)

3.2P: Exercise 3.2 (for mathematicians)

3.3P: Consider the two lines g_1 and g_2 in $\mathbb{R}^3$ described by

$$g_i := \{p_i + tv_i \mid t \in \mathbb{R}\}, \quad i = 1, 2$$

where

$$\begin{aligned} p_1 &:= (1, 1, 2) \\ p_2 &:= (0, -1, 3) \\ v_1 &:= (2, 0, 1) \\ v_2 &:= (1, 1, 1) \end{aligned}$$

How large is the distance a between g_1 and g_2?

This exercise has to do with the vector product, for if $q_1 \in g_1$ and $q_2 \in g_2$ are the two points on the lines at the shortest distance apart, i.e., $\|q_1 - q_2\| = a$, then $q_1 - q_2 \in \mathbb{R}^3$ is perpendicular to both lines, that is, to their directions v_1 and v_2. (To check your solution: the third and fourth decimal places of a should be 1 and 2.)

Chapter 4
Linear Maps

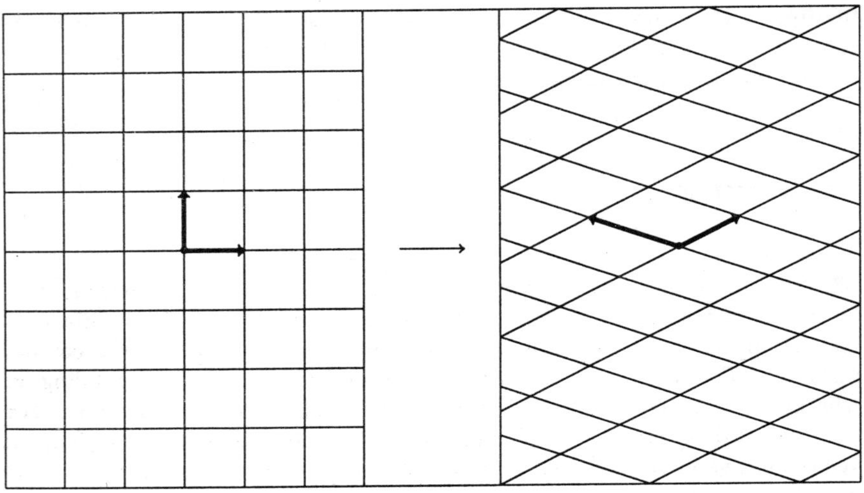

4.1 Linear Maps

Until now we have always studied a single vector space V and various objects in it, like r-tuples of linearly independent vectors, subspaces, bases, and so forth. Now we want to consider *two* vector spaces V and W and study relations between what is going on in V and W respectively. Such relations will be described by "linear maps" or "homomorphisms." A map $f : V \to W$ is called linear if it is compatible with the vector space operations $+$ and $\cdot$ in V and W, that is, if it is irrelevant whether I first add two elements in V and then map the sum, or if I first map them and afterwards add their images, and similarly for scalar multiplication.

Definition: Let V and W be vector spaces over $\mathbb{F}$. A map $f : V \to W$ is called ***linear***, or a ***homomorphism***, if for all $x, y \in V$, $\lambda \in \mathbb{F}$ we have

$$f(x + y) = f(x) + f(y)$$
$$f(\lambda x) = \lambda f(x).$$

The set of homomorphisms from V to W is denoted by $\mathrm{Hom}(V, W)$.

Fact 1: The identity $\mathrm{Id}_V : V \to V$ is linear, and if $V \xrightarrow{f} W \xrightarrow{g} Y$ are linear maps, then $gf : V \to Y$ is also a linear map.

Fact 2: If for all $f, g \in \mathrm{Hom}(V, W)$ and $\lambda \in \mathbb{F}$, the elements $f + g$ and λf in $\mathrm{Hom}(V, W)$ are defined in the obvious way, then with these two operations $\mathrm{Hom}(V, W)$ is a vector space over $\mathbb{F}$.

The "obvious" or "canonical way" consists in defining $(f + g)(x)$ to be $f(x) + g(x)$ and $(\lambda f)(x)$ to be $\lambda f(x)$.

For each linear map $f : V \to W$ two subspaces of V and W, respectively, are particularly important. The first is the image $f(V)$ of V under f, or the "image of f" for short, which is a subspace of W. The other is called the "kernel of f," which is the subspace $f^{-1}(0) = \{v \in V \mid f(v) = 0\}$ of V. See Fig. 28 for an example. They really *are* subspaces, as one sees immediately from the definitions of subspace (in Section 2.3) and linear map.

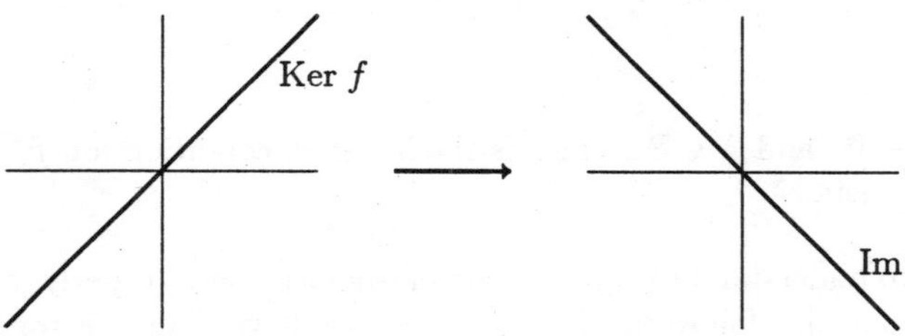

Fig. 28. Kernel and image of the linear map $f: \mathbb{R}^2 \to \mathbb{R}^2, (x,y) \mapsto (x-y, y-x)$

Fact 3 and Definition: Let $f : V \to W$ be a linear map. Then the image $\mathrm{Im}\, f := f(V)$ of f is a subspace of W, and the ***kernel***

$$\mathrm{Ker}\, f := \{v \in V \mid f(v) = 0\}$$

of f is a subspace of V. The linear map f is injective if and only if $\mathrm{Ker}\, f = 0$, because $f(x) = f(y)$ means $x - y \in \mathrm{Ker}\, f$.

Linear maps with certain particular properties have almost self-explanatory special names.

Definition: A linear map $f : V \to W$ is called
a ***monomorphism*** if it is injective,
an ***epimorphism*** if it is surjective,
an ***isomorphism*** if it is bijective,
an ***endomorphism*** if $V = W$, and finally
an ***automorphism*** if it is bijective and $V = W$.

The *isomorphisms* are particularly important. That the composition $g \circ f$ of two isomorphisms $f : V \to W$ and $g : W \to Y$ is again an isomorphism is clear from Fact 1. However, the following is worth noting.

> **Remark 1:** If $f : V \to W$ is an isomorphism, then $f^{-1} : W \to V$ is also an isomorphism.

PROOF: Since f^{-1} is again bijective, we must convince ourselves only that f^{-1} is also linear. It follows immediately from the definition of a linear map that for all $x, y \in V, \lambda \in \mathbb{F}$, we have

$$f^{-1}(f(x+y)) = f^{-1}(f(x) + f(y))$$
$$f^{-1}(f(\lambda x)) = f^{-1}(\lambda f(x)).$$

If $v, w \in W$ and we write $f^{-1}(v) = x$, $f^{-1}(w) = y$, then this gives

$$f^{-1}(v) + f^{-1}(w) = f^{-1}(v+w)$$
$$\lambda f^{-1}(v) = f^{-1}(\lambda v)$$

for all $v, w \in W$ and $\lambda \in \mathbb{F}$, which is the linearity condition for f^{-1} — read from right to left. □

In order to understand the importance of isomorphisms properly, you should be clear about the following. Assume that we have some vector space V containing various objects — subsets, subspaces, bases and the like. If now $\varphi : V \to W$ is an isomorphism, we can consider the images of our "objects" in W (see Fig. 29).

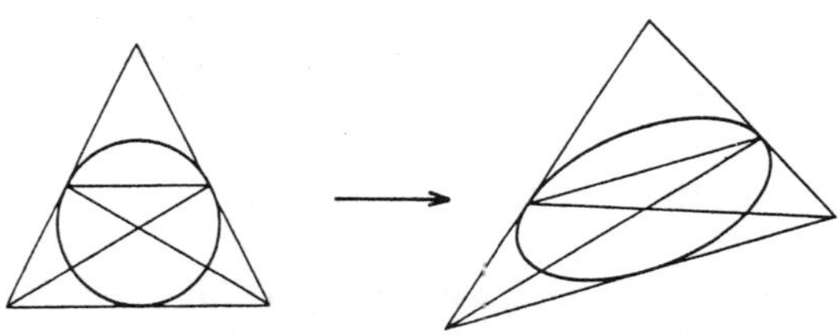

Fig. 29. Isomorphism φ applied to some substructure of V

These images in W then have the same linear properties as the original objects in V. Somewhat vaguely put, "linear properties" are those which can be formulated in terms of the vector space data *sets, addition, and scalar multiplication.* Example: suppose that U_1, U_2 are two subspaces of V, and that $U_1 \cap U_2$ has dimension five. Then the subspace $\varphi(U_1) \cap \varphi(U_2)$ of W also has dimension five. Again: if $(v_1, \ldots, v_r)$ is a linearly independent r-tuple of vectors in V, then $(\varphi(v_1), \ldots, \varphi(v_r))$ is also a linearly independent r-tuple of vectors in W.

Examples of nonlinear properties: suppose first that $V = \mathbb{R}^2$. Then each $x \in V$ is a pair of numbers. If $\varphi : V \to W$ is an isomorphism, $\varphi(x)$ need not be a pair of numbers; W may perhaps be a vector space whose elements are *functions* or the like. Next, suppose that $V = W = \mathbb{R}^2$. Let $U \subset \mathbb{R}^2$ be a circle: $U = \{(x_1, x_2) \in \mathbb{R}^2 \mid x_1^2 + x_2^2 = 1\}$. If $\varphi : \mathbb{R}^2 \to \mathbb{R}^2$ is an isomorphism, it does not necessarily follow that $\varphi(U) \subset \mathbb{R}^2$ is a circle (see Fig. 30). Thus $\mathbb{R}^2 \to \mathbb{R}^2$, $(x_1, x_2) \mapsto (2x_1, x_2)$ is such an isomorphism.

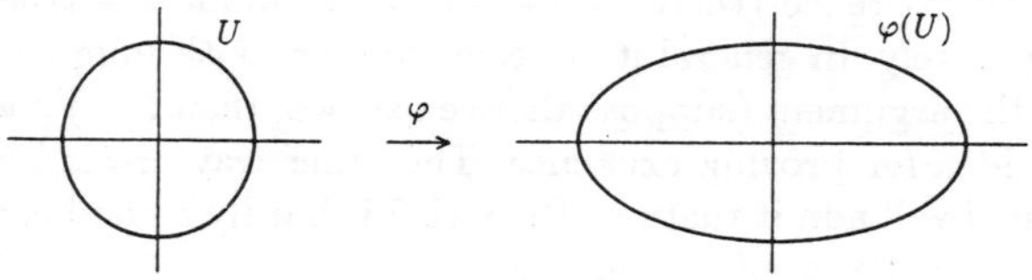

Fig. 30. To be a circle is not a "linear" property

I will not try to be more formal about the concept of a linear property. In time you will come to know many examples of properties that are "invariant under isomorphism."

Another very important aspect of isomorphisms between vector spaces is their use in relating *linear maps* with one another. Imagine for a moment that we are interested in a particular linear map $f : V \to V$, which, however, is at first difficult to handle — possibly because V is a function space and f is a complicated differential or integral operator from analysis. Imagine further that we have some specific "linear questions" about the map f. For example, is it injective or surjective, how big is the dimension of the image $f(V)$, do there exist vectors $v \neq 0$ in V mapped to a multiple λv of themselves by f (called *eigenvectors*), and the like.

Now in this situation it is sometimes possible to find some other vector space V' and an isomorphism $\varphi : V' \cong V$, which turns f into an easily understandable map $f' := \varphi^{-1} \circ f \circ \varphi$,

$$\begin{array}{ccc} V & \xrightarrow{\ f\ } & V \\ {\scriptstyle\varphi}\big\uparrow{\scriptstyle\cong} & & {\scriptstyle\varphi}\big\uparrow{\scriptstyle\cong} \\ V' & \xrightarrow[\ f'\ ]{} & V' \end{array}$$

for which we can immediately answer the analogous questions. These answers can then be carried back by means of φ to f, in which we are really interested. For example: if $v' \in V'$ is an eigenvector of f' with $f'(v') = \lambda v'$, then for $v = \varphi(v')$ we also have $f(v) = \lambda v$, and so forth.

With this we end our little digression on the importance of the concept of isomorphism, and return again to the details of Chapter 4.

We now want to assume that V is finite-dimensional and to note some general statements about linear maps $f : V \to W$ connected with this. Whoever until now has felt a lack of examples of linear maps will find the following remark richly rewarding.

Remark 2: Let V and W be vector spaces over $\mathbb{F}$ and $(v_1, \ldots, v_n)$ be a basis of V. Then for each n-tuple $(w_1, \ldots, w_n)$ of vectors in W there exists a unique linear map $f : V \to W$ with $f(v_i) = w_i$, $i = 1, \ldots, n$.

PROOF: For such "there exists a unique" statements it mostly happens that one must prove existence (there exists one) and uniqueness (there exists at most one) separately. In general it is better to start with uniqueness, since in the course of the argument (suppose there exist two, then . . .) one sometimes arrives at an idea for proving existence. The other way around happens less frequently. But I will admit that our Remark 2 is hardly a good example, since here both parts of the proof are very easy.

(a) PROOF OF UNIQUENESS: Suppose that $f, f' : V \to W$ are linear maps with $f(v_i) = f'(v_i) = w_i$, $i = 1, \ldots, n$. Then since each $v \in V$ can be written as $v = \lambda_1 v_1 + \cdots + \lambda_n v_n$, we have

$$\begin{aligned} f(v) &= f(\lambda_1 v_1 + \cdots + \lambda_n v_n) \\ &= \lambda_1 f(v_1) + \cdots + \lambda_n f(v_n) \\ &= \lambda_1 f'(v_1) + \cdots + \lambda_n f'(v_n) \\ &= f'(\lambda_1 v_1 + \cdots + \lambda_n v_n) = f'(v). \end{aligned}$$

Hence $f(v) = f'(v)$ for all $v \in V$.

(b): PROOF OF EXISTENCE: Since each $v \in V$ can *uniquely* be written as $v = \lambda_1 v_1 + \cdots + \lambda_n v_n$, we genuinely define a map $f : V \to W$ by letting

$$f(\lambda_1 v_1 + \cdots + \lambda_n v_n) := \lambda_1 w_1 + \cdots + \lambda_n w_n .$$

Clearly, f is linear and has the property $f(v_i) = w_i$, $i = 1, \ldots, n$. □

This innocent-looking and easy-to-prove remark has nevertheless a very important content: *complete information about a linear map is contained in the images of the basis vectors!* Take $V = \mathbb{F}^n$ with its canonical basis and $W = \mathbb{F}^m$ as an example. By Remark 2, to give a linear map $\mathbb{F}^n \to \mathbb{F}^m$ is the same as giving m-tuples $w_1, \ldots, w_n \in \mathbb{F}^m$, amounting to altogether $n \cdot m$ numbers from $\mathbb{F}$, in which the linear map is then encoded. This is the reason why one can carry out effective computer calculations for linear maps, and why one always tries to use theoretical considerations to reduce nonlinear problems to linear ones.

Remark 3: Let V and W be vector spaces over $\mathbb{F}$ and let $(v_1, \ldots, v_n)$ be a basis of V. A linear map $f : V \to W$ is an isomorphism, if and only if $(f(v_1), \ldots, f(v_n))$ is a basis of W.

PROOF: If you first recall the definitions of "basis" and "isomorphism," then place the tip of your ball-point pen on the paper and move your hand a little — the proof will flow out effortlessly. The terminology thinks for you! Sooner or later we must stop writing out such "proofs" every time. Just look at this one:

(a): Let f be an isomorphism. We first prove the linear independence of $(f(v_1), \dots, f(v_n))$ in W. So let $\lambda_1 f(v_1) + \dots + \lambda_n f(v_n) = 0$. This means $f(\lambda_1 v_1 + \dots + \lambda_n v_n) = 0$ because of the linearity of f. Since f is injective and $f(0) = 0$, we must have $\lambda_1 v_1 + \dots + \lambda_n v_n = 0$. Since $(v_1, \dots, v_n)$ is linearly independent, it follows from this that $\lambda_1 = \dots = \lambda_n = 0$, and hence that $(f(v_1), \dots, f(v_n))$ is linearly independent.

Now we show that $L(f(v_1), \dots, f(v_n)) = W$. Let $w \in W$. Since f is surjective, there exists $v \in V$ with $f(v) = w$. Since $L(v_1, \dots, v_n) = V$, there exist $\lambda_1, \dots, \lambda_n \in \mathbb{F}$ with $v = \lambda_1 v_1 + \dots + \lambda_n v_n$. Since f is linear, we have $w = f(v) = \lambda_1 f(v_1) + \dots + \lambda_n f(v_n)$, and so every element of W can be written as a linear combination of $f(v_1), \dots, f(v_n)$.

(b): Let $(f(v_1), \dots, f(v_n))$ be a basis of W. First we prove the injectivity of f. Let $f(v) = 0$. Since $(v_1, \dots, v_n)$ is a basis of V, there exist $\lambda_1, \dots, \lambda_n \in \mathbb{F}$ with $v = \lambda_1 v_1 + \dots + \lambda_n v_n$. Then, because of the linearity of f, we also have that $\lambda_1 f(v_1) + \dots + \lambda_n f(v_n) = 0$, and because $(f(v_1), \dots, f(v_n))$ is linearly independent, it follows from this that $\lambda_1 = \dots = \lambda_n = 0$. Hence $v = 0$ and f is injective.

Now we prove the surjectivity of f. Let $w \in W$. Since $f(v_1), \dots, f(v_n)$ generate all of W, there exist $\lambda_1, \dots, \lambda_n$ with $w = \lambda_1 f(v_1) + \dots + \lambda_n f(v_n)$. Let $v := \lambda_1 v_1 + \dots + \lambda_n v_n$. Then, because of linearity, $f(v) = \lambda_1 f(v_1) + \dots + \lambda_n f(v_n)$ $= w$, and f is surjective. □

Remarks 2 and 3 together have the following consequence.

Fact 4: Any two n-dimensional vector spaces over $\mathbb{F}$ are isomorphic. (To say that V and W are isomorphic means of course that there is an isomorphism $f : V \cong W$.)

This is also very remarkable. "Up to isomorphism," as one says, there exists only *one* n-dimensional vector space over $\mathbb{F}$! Nevertheless, it would be unwise to study $\mathbb{F}^n$ alone, since quite uninvited, all sorts of other concrete vector spaces tumble across our path (solution spaces, function spaces, tangent spaces, etc.), and in order to understand them and their relation with $\mathbb{F}^n$, we need the general concept of a vector space.

Our last topic in this section will be a dimension formula. When in linear algebra one has to deal with several vector spaces at the same time, it is often very useful to have a formula giving a relation between the dimensions of the individual spaces. For example, in Chapter 3 we proved such a dimension formula for subspaces U_1, U_2 of V. Thus $\dim(U_1 \cap U_2) + \dim(U_1 + U_2) = \dim U_1 + \dim U_2$. Now we want to obtain a dimension formula concerning linear maps.

Definition: Let $f : V \to W$ be a linear map. If the image $\operatorname{Im} f$ is finite-dimensional, $\operatorname{rk} f := \dim \operatorname{Im} f$ is called the ***rank*** of f.

Dimension formula for linear maps: Let V be a finite-dimensional vector space and $f: V \to W$ a linear map. Then

$$\dim \operatorname{Ker} f + \operatorname{rk} f = \dim V.$$

PROOF: Let n be the dimension of V and r the dimension of the kernel. We extend a basis $(v_1, \ldots, v_r)$ of $\operatorname{Ker} f$ to a basis $(v_1, \ldots, v_r, v_{r+1}, \ldots, v_n)$ of all of V and put $w_i := f(v_{r+i})$ for $i = 1, \ldots, n-r$. Then we have

$$f(\lambda_1 v_1 + \cdots + \lambda_n v_n) = \lambda_{r+1} w_1 + \cdots + \lambda_n w_{n-r},$$

and so the image of f is $L(w_1, \ldots, w_{n-r})$. Moreover $(w_1, \ldots, w_{n-r})$ is linearly independent, since from $\alpha_1 w_1 + \cdots + \alpha_{n-r} w_{n-r} = 0$, say, it would follow that $\alpha_1 v_{r+1} + \cdots + \alpha_{n-r} v_n \in \operatorname{Ker} f$. Hence we would have $\alpha_1 v_{r+1} + \cdots + \alpha_{n-r} v_n = \mu_1 v_1 + \cdots + \mu_r v_r$ for suitable $\mu_1, \ldots, \mu_r$, but $(v_1, \ldots, v_n)$ is linearly independent and therefore all the α's and μ's would have to vanish. Thus $(w_1, \ldots, w_{n-r})$ is indeed a basis of the image of f, and hence $\dim \operatorname{Im} f = n - r$. □

As an application of the dimension formula, let us note the following fact.

Fact 5: A linear map between two spaces of the same dimension is surjective if and only if it is injective.

4.2 Matrices

Definition: An $m \times n$ ***matrix*** over $\mathbb{F}$ is an array of mn elements from $\mathbb{F}$ according to the following pattern

$$\begin{pmatrix} a_{11} & \cdots\cdots\cdots & a_{1n} \\ \vdots & & \vdots \\ a_{m1} & \cdots\cdots\cdots & a_{mn} \end{pmatrix}.$$

The $a_{ij} \in \mathbb{F}$ are called the ***coefficients*** of the matrix. The horizontally written n-tuples $(a_{i1} \ldots a_{in})$ are called the ***rows***, and the vertically written m-tuples

$$\begin{pmatrix} a_{1j} \\ \vdots \\ a_{mj} \end{pmatrix}$$

are called the ***columns*** of the matrix (see Fig. 31). The set of all $m \times n$ matrices over $\mathbb{F}$ is denoted by $M(m \times n, \mathbb{F})$.

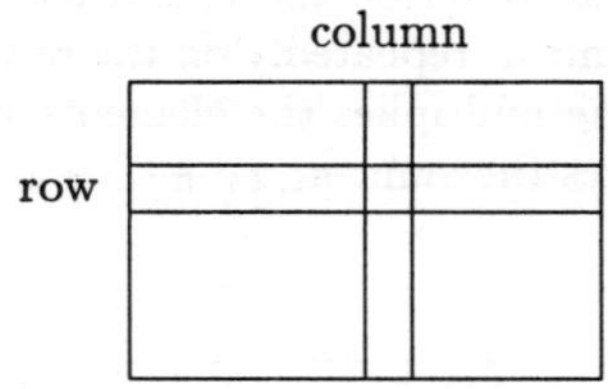

Fig. 31. Rows and columns in a matrix

Matrices play a role in various contexts in linear algebra. For the moment tney are of interest to us because of their importance for linear maps.

Definition: For $x = (x_1, \dots, x_n) \in \mathbb{F}^n$ define $Ax \in \mathbb{F}^m$ by

$$Ax = (\sum_{i=1}^{n} a_{1i}x_i, \ \sum_{i=1}^{n} a_{2i}x_i, \dots, \sum_{i=1}^{n} a_{mi}x_i).$$

There is another very suggestive way of writing this "operation" of a matrix A on an element $x \in \mathbb{F}^n$.

Notation: In connection with the operation of $m \times n$ matrices on n-tuples it is customary to write the elements of $\mathbb{F}^n$ and $\mathbb{F}^m$ as columns:

$$\begin{pmatrix} a_{11} & \cdots\cdots & a_{1n} \\ \vdots & & \vdots \\ a_{m1} & \cdots\cdots & a_{mn} \end{pmatrix} \begin{pmatrix} x_1 \\ \vdots \\ x_n \end{pmatrix} = \begin{pmatrix} a_{11}x_1 + \cdots + a_{1n}x_n \\ \vdots \\ a_{m1}x_1 + \cdots + a_{mn}x_n \end{pmatrix}$$

Please look at this closely: in spite of its "rectangular" appearence the right-hand side is no $m \times n$ matrix, but only an m-tuple written as a column.

Before we come to the mathematics of the relation between matrices and linear maps, let us make one further remark about the handling of the numerous indices. Formulae, like that above for Ax with many indices, can be more easily remembered if one has certain mnemonics for them. The picture in Fig. 32 provides such a mnemonic.

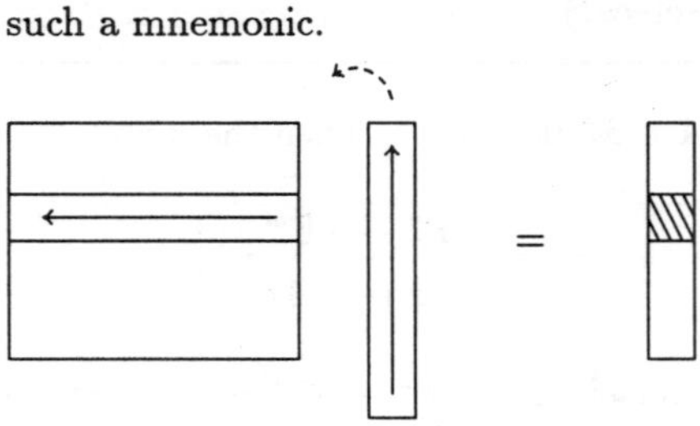

Fig. 32. Calculating the ith component of Ax

More explicitly, one can describe the construction of the column Ax as follows. One lays the column x repeatedly on the rows of A by giving it a 90^0 turn like a stick. Then one multiplies the elements a_{ij} and x_j now lying on top of each other and takes the sum: $a_{i1}x_1 + \cdots + a_{in}x_n$.

Construction of the first

second

...

...

last component of Ax

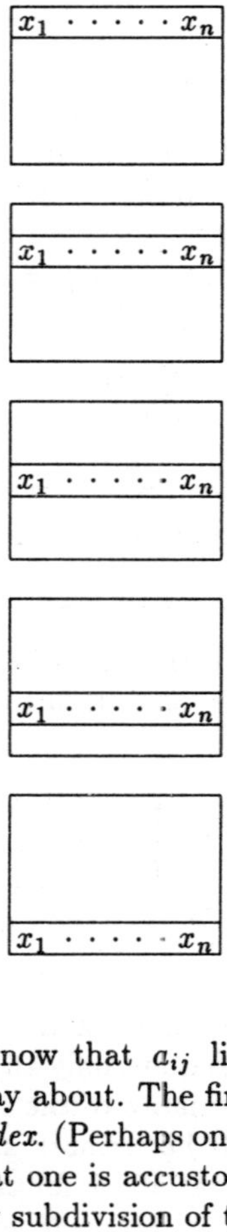

Of course, here one has to know that a_{ij} lies in the ith row and the jth column, and not the other way about. The first index is called the *row index* and the second the *column index*. (Perhaps one can remember this designation of the indices by thinking that one is accustomed to read in rows, and hence that the rows, as the primary subdivision of the matrix, can lay claim to the first index for themselves?)

Theorem: Let $A \in M(m \times n, \mathbb{F})$. Then the map

$$\begin{aligned} \mathbb{F}^n &\longrightarrow \mathbb{F}^m \\ x &\longmapsto Ax \end{aligned}$$

is linear, and conversely, if $f : \mathbb{F}^n \to \mathbb{F}^m$ is a linear map, there exists a unique matrix $A \in M(m \times n, \mathbb{F})$ with $f(x) = Ax$ for all $x \in \mathbb{F}^n$.

The association of a matrix with its corresponding linear map thus defines a bijective map $M(m \times n, \mathbb{F}) \to \mathrm{Hom}(\mathbb{F}^n, \mathbb{F}^m)$! Hence one can *interpret* or *consider* the $m \times n$ matrices as the linear maps from $\mathbb{F}^n$ to $\mathbb{F}^m$.

PROOF OF THE THEOREM: One immediately reads from the definition of Ax that $A(x+y) = Ax + Ay$, and that $A(\lambda x) = \lambda(Ax)$ for all $x, y \in \mathbb{F}^n$ and $\lambda \in \mathbb{F}$. The map given by $x \mapsto Ax$ is therefore linear. Now let $f : \mathbb{F}^n \to \mathbb{F}^m$ be an arbitrary linear map. We must show that there exists a unique matrix $A \in M(m \times n, \mathbb{F})$, such that $f(x) = Ax$ for all x in $\mathbb{F}^n$. We again divide this "there exists a unique" proof between existence and uniqueness.

(a) PROOF OF UNIQUENESS: Suppose that $A, B \in M(m \times n, \mathbb{F})$ and that $f(x) = Ax = Bx$ for all $x \in \mathbb{F}^n$. Then in particular for the "unit vectors" e_i, that is for

$$e_1 = \begin{pmatrix} 1 \\ 0 \\ \vdots \\ 0 \end{pmatrix}, \quad \ldots, \quad e_n = \begin{pmatrix} 0 \\ \vdots \\ 0 \\ 1 \end{pmatrix} \in \mathbb{F}^n,$$

we have $Ae_i = Be_i$, $i = 1, \ldots, n$. But what is Ae_i? Ae_i is just the ith column of A, as Fig. 33 shows. Therefore A and B have the same columns and so are equal.

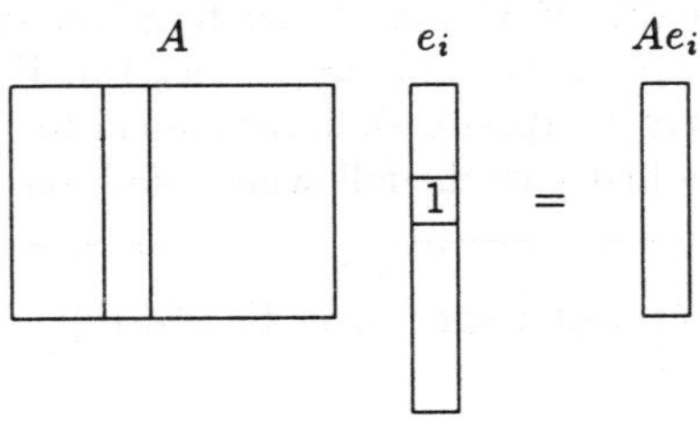

Fig. 33. Why Ae_i is just the ith column of A

Let us write this out more formally. Let $\delta_{i1}, \ldots, \delta_{in}$ denote the components of e_i, thus

$$\delta_{ij} = \begin{cases} 1 & \text{for } i = j \\ 0 & \text{for } i \neq j. \end{cases}$$

Then the kth component of Ae_i, written as $(Ae_i)_k$, is given by

$$(Ae_i)_k = \sum_{j=1}^{n} a_{kj}\delta_{ij} = a_{ki},$$

and we have $a_{ki} = (Ae_i)_k = (Be_i)_k = b_{ki}$ for all $i = 1, \ldots, n$ and $k = 1, \ldots, m$. Therefore $A = B$.

(b) PROOF OF EXISTENCE: For each $m \times n$ matrix A, as we have just seen from the uniqueness proof, the columns of the matrix are the images of the unit vectors $e_i \in \mathbb{F}^n$ for the map $\mathbb{F}^n \to \mathbb{F}^m$, $x \mapsto Ax$. *"The columns are the images of the unit vectors"* is in any case useful motto in matrix arithmetic.

Since we want to have $Ax = f(x)$ for all x, in particular for $x = e_i$, it follows that we *must* define A so that

$$f(e_i) =: v_i =: \begin{pmatrix} v_{1i} \\ \vdots \\ v_{mi} \end{pmatrix} \in \mathbb{F}^m$$

becomes the ith column. We therefore write

$$A := \begin{pmatrix} v_{11} & \cdots\cdots\cdots & v_{1n} \\ \vdots & & \vdots \\ v_{m1} & \cdots\cdots\cdots & v_{mn} \end{pmatrix}$$

and hope for the best. At least we have obtained a matrix for which $Ae_i = f(e_i)$ for all $i = 1, \ldots, n$. But because of the linearity of both f and the map given by $x \mapsto Ax$, it follows that $A(\lambda_1 e_1 + \cdots + \lambda_n e_n) = f(\lambda_1 e_1 + \cdots + \lambda_n e_n)$ for arbitrary $\lambda_i \in \mathbb{F}$, and since $(e_1, \ldots, e_n)$ is a basis of $\mathbb{F}^n$, this means that $Ax = f(x)$ for all $x \in \mathbb{F}^n$. □

What has all this to do with the linear maps from a vector space V into a vector space W? Just this: if V and W are finite-dimensional vector spaces and we choose bases $(v_1, \ldots, v_n)$ and $(w_1, \ldots, w_m)$ in V and W respectively, then we can immediately interpret each linear map in the language of matrices. To explain this, let me first give the following definition.

Definition: If V is a vector space over $\mathbb{F}$, and $(v_1, \ldots, v_n)$ is a basis of V, we say that the map

$$\begin{aligned} K^n &\xrightarrow{\cong} V \\ (\lambda_1, \ldots, \lambda_n) &\longmapsto \lambda_1 v_1 + \cdots + \lambda_n v_n \end{aligned}$$

is the ***canonical basis isomorphism***. If some notation is necessary we write $\Phi_{(v_1 \ldots v_n)}$ for this isomorphism.

By Remarks 2 and 3 the basis isomorphism is precisely the uniquely determined isomorphism mapping the unit vectors in $\mathbb{F}^n$ to the vectors of the given basis. If now $(v_1, \ldots, v_n)$ and $(w_1, \ldots, w_m)$ are bases of V and W, respectively, and $f : V \to W$ is a linear map, then

$$\Phi^{-1}_{(w_1, \ldots, w_m)} \circ f \circ \Phi_{(v_1, \ldots, v_n)}$$

is a linear map from $\mathbb{F}^n$ to $\mathbb{F}^m$ and hence given by an $m \times n$ matrix A. A commutative diagram is the clearest way to describe the connection between f and A.

Definition: Let $f : V \to W$ be a linear map between vector spaces over $\mathbb{F}$, and let $(v_1, \ldots, v_n)$ and $(w_1, \ldots, w_m)$ be bases for V and W, repectively. Then the matrix $A \in M(m \times n, \mathbb{F})$ determined by the commutative diagram

$$\begin{array}{ccc} V & \xrightarrow{\ f\ } & W \\ \Phi_{(v_1,\ldots,v_n)} \Big\uparrow \cong & & \cong \Big\uparrow \Phi_{(w_1,\ldots,w_m)} \\ \mathbb{F}^n & \xrightarrow{\ A\ } & \mathbb{F}^m \end{array}$$

is called the matrix ***associated*** to f relative to the two chosen bases.

Hence if one has to consider linear maps between finite-dimensional vector spaces, one can always pass to the associated matrices by choice of bases. Conversely, one can always reconstruct f from the matrix A, since we have

$$\Phi_{(w_1,\ldots,w_m)} \circ A \circ \Phi^{-1}_{(v_1,\ldots,v_n)} = f.$$

In particular, the correspondence between $\mathrm{Hom}(V, W)$ and $M(m \times n, \mathbb{F})$ given by $f \mapsto A$ and determined by choice of bases is bijective.

Passage to matrices is not only useful in the carrying out of concrete calculations, but in the right circumstances it can even help with theoretical considerations. We must realize one thing, however: *in altering the bases one also alters the matrix that represents f*. This is sometimes a blessing, since by a cunning choice of bases one can arrive at very simple matrices, and is also sometimes a curse, since the close observation of alterations brought about by change of bases ("behavior under transformations") can be both necessary and tiresome.

4.3 Test

(1) A map $f : V \to W$ between vector spaces V and W over $\mathbb{F}$ is linear, if

☐ $f(\lambda x + \mu y) = \lambda f(x) + \mu f(y)$ for all $x, y \in V$, $\lambda, \mu \in \mathbb{F}$.
☐ f satisfies the eight axioms for a vector space.
☐ $f : V \to W$ is bijective.

(2) By the kernel of a linear map $f : V \to W$ one understands

☐ $\{w \in W \mid f(0) = w\}$
☐ $\{f(v) \mid v = 0\}$
☐ $\{v \in V \mid f(v) = 0\}$

(3) Which of the following statements are correct? If $f : V \to W$ is a linear map, we have

☐ $f(0) = 0$.
☐ $f(-x) = -x$ for all $x \in V$.
☐ $f(\lambda v) = f(\lambda) + f(v)$ for all $\lambda \in \mathbb{F}$, $v \in V$.

(4) A linear map $f : V \to W$ is called an isomorphism if

☐ there exists a linear map $g : W \to V$ with $fg = \mathrm{Id}_W$ and $gf = \mathrm{Id}_V$.
☐ V and W are isomorphic.
☐ for each n-tuple $(v_1, \ldots, v_n)$ in V, the n-tuple $(f(v_1), \ldots, f(v_n))$ is a basis of W .

(5) By the rank $\mathrm{rk}(f)$ of a linear map $f : V \to W$, one understands

☐ $\dim \operatorname{Ker} f$ ☐ $\dim \operatorname{Im} f$ ☐ $\dim W$

(6) $\begin{pmatrix} 1 & -1 \\ 2 & 0 \end{pmatrix} \begin{pmatrix} 3 \\ 1 \end{pmatrix} =$

☐ $\begin{pmatrix} 2 \\ 6 \end{pmatrix}$ ☐ $\begin{pmatrix} 5 \\ -3 \end{pmatrix}$ ☐ $\begin{pmatrix} 0 \\ 2 \end{pmatrix}$

(7) The map $\mathbb{R}^2 \to \mathbb{R}^2$, $(x, y) \mapsto (x + y, x - y)$, is given by the following matrix ("The columns are the ..."):

☐ $\begin{pmatrix} 1 & 1 \\ 1 & -1 \end{pmatrix}$ ☐ $\begin{pmatrix} 0 & 2 \\ -2 & 0 \end{pmatrix}$ ☐ $\begin{pmatrix} 1 & 1 \\ -1 & 1 \end{pmatrix}$

(8) Let V and W be two vector spaces with bases (v_1, v_2, v_3) and (w_1, w_2, w_3) and let $f : V \to W$ be the linear map with $f(v_i) = w_i$. Then the "associated" matrix is

☐ $A = \begin{pmatrix} 1 & 1 & 1 \\ 1 & 1 & 1 \\ 1 & 1 & 1 \end{pmatrix}$ ☐ $A = \begin{pmatrix} 1 & 0 & 0 \\ 0 & 1 & 0 \\ 0 & 0 & 1 \end{pmatrix}$ ☐ $A = \begin{pmatrix} 0 & 0 & 0 \\ 0 & 0 & 0 \\ 0 & 0 & 0 \end{pmatrix}$

(9) A linear map $f : V \to W$ is injective if and only if

☐ f is surjective. ☐ $\dim \operatorname{Ker} f = 0$. ☐ $\mathrm{rk} f = 0$.

(10) Let $f : V \longrightarrow W$ be a surjective linear map and $\dim V = 5$, $\dim W = 3$. Then

☐ $\dim \operatorname{Ker} f \geq 3$.
☐ $\dim \operatorname{Ker} f$ is 0, 1, or 2 and each of these cases can arise.
☐ $\dim \operatorname{Ker} f = 2$.

4.4 Quotient Spaces

A section for mathematicians

Let V be a vector space over a field $\mathbb{F}$ and let $U \subset V$ be a subspace. We want to define the quotient "V over U," or "V mod U," and this will again be a vector space over $\mathbb{F}$. We know that a vector space consists of three things: a set, addition, and scalar multiplication. In our case the set will be denoted by V/U, and hence we have to define the triple:

(a) the set V/U,
(b) the addition $V/U \times V/U \to V/U$,
(c) the scalar multiplication $\mathbb{F} \times V/U \to V/U$.

Then, of course, we still have to check the eight axioms.

FOR (a): If $x \in V$, let $x + U := \{x + u \mid u \in U\}$, and define V/U to be the set of all $x + U$, i.e.,

$$V/U := \{x + U \mid x \in V\}.$$

Note that the elements of V/U are not those elements of V that have the form $x + u$ with $u \in U$, but each element of V/U is itself a *set* $\{x + u \mid u \in U\}$. Look at Fig. 34.

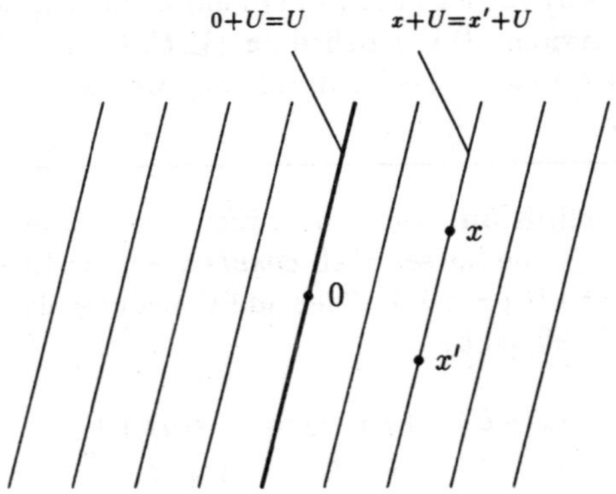

Fig. 34. The elements $x+U$ of the quotient space are certain subsets of V.

Of course, U itself is also an element of V/U, namely $0 + U$.

FOR (b): We wish to define addition in V/U by $(x+U)+(y+U) := (x+y)+U$. A strange difficulty appears here. Although we can write down this obvious formula for the sum, we have to ask ourselves whether in this way $(x+U)+(y+U)$ is actually *well defined*, because it can happen that $x + U = x' + U$ even though $x \neq x'$! Only when we have shown that with $x + U = x' + U$ and $y + U = y' + U$ we also have $(x + y) + U = (x' + y') + U$, are we allowed to say that $(x + U) + (y + U) := (x + y) + U$ really *defines* a composition $V/U \times V/U \to V/U$.

So let us check this. Assume $x + U = x' + U$ and $y + U = y' + U$. Then in particular, x itself, as $x + 0 \in x + U$, is in $x' + U$ and y in $y' + U$, which means that there exist $a, b \in U$ with $x = x' + a$ and $y = y' + b$, and hence

$$\begin{aligned}(x + y) + U &= (x' + a) + (y' + b) + U \\ &= ((x' + y') + (a + b)) + U \\ &= \{x' + y' + a + b + u \mid u \in U\}.\end{aligned}$$

But $u \in U$ implies $u' := a + b + u \in U$ (since $a, b \in U$) and, conversely, *each* element $u' \in U$ can be written as $a + b + u$ for some $u \in U$, namely for $u = u' - (a + b)$. Hence

$$\begin{aligned}(x + y) + U &= \{x' + y' + (a + b + u) \mid u \in U\} \\ &= \{x' + y' + u' \mid u' \in U\} \\ &= (x' + y') + U.\end{aligned}$$

FOR (c): Here again we must watch out for our operation being "well defined," if we put $\lambda(x + U) := \lambda x + U$. If $x = x' + a$ with $a \in U$, we have $\lambda x + U = \lambda x' + \lambda a + U = \lambda x' + U$, because U is a subspace. Hence the scalar multiplication $\lambda(x + U) := \lambda x + U$ is well defined.

Are the eight vector space axioms satisfied for $(V/U, +, \cdot)$? The validity of axioms (1), (2), and (5)–(8) for $(V/U, +, \cdot)$ follows immediately from their validity for $(V, +, \cdot)$. Axiom (3) is satisfied for $(V/U, +, \cdot)$ with $0 := U \in V/U$, and (4) with $-(x + U) := (-x) + U$. In this way we see that V/U is indeed a vector space over $\mathbb{F}$.

Remark and Definition: Let V be a vector space over $\mathbb{F}$ and $U \subset V$ a subspace. Let V/U be the set of all ***cosets*** $x + U := \{x + u \mid u \in U\}$ of U, thus $V/U = \{x + U \mid x \in V\}$. Then addition and scalar multiplication for V/U are well defined by

$$\begin{aligned}(x + U) + (y + U) &:= (x + y) + U \\ \lambda(x + U) &:= \lambda x + U\end{aligned}$$

for all $x, y \in V, \lambda \in \mathbb{F}$. These operations make V/U into a vector space over $\mathbb{F}$, which is called the ***quotient vector space*** of V modulo U.

PROOF: Since U is a subspace of V, we have $x + U = x' + U \Leftrightarrow x - x' \in U$. This easily implies that the operations are well defined. If one writes $0 := U \in V/U$ and $-(x + U) := (-x) + U$, the validity of the eight vector space axioms for V/U follows from the validity of the corresponding axioms for V. Hence $(V/U, +, \cdot)$ is a vector space. □

It follows immediately from the definition that V/V consists of a single element, since $x + V = V$ for all $x \in V$. Therefore V/V is the vector space consisting of zero only. At the opposite extreme there is only a formal difference between $V/\{0\}$ and V: the obvious map $v \mapsto \{v\}$ defines an isomorphism from V to $V/\{0\}$.

In order not to have to return to the definition in all arguments, it is useful to remember two or three basic properties of quotient spaces.

Fact 1: The map

$$\begin{aligned} \pi : V &\longrightarrow V/U \\ v &\longmapsto v + U \end{aligned}$$

is an epimorphism with $\operatorname{Ker} \pi = U$. (The map π is called the ***projection.***)

Fact 2: If V is finite-dimensional, we have

$$\dim V/U = \dim V - \dim U.$$

Lemma: If $f : V \to W$ is a linear map with $U \subset \operatorname{Ker} f$, there exists a unique linear map $\varphi : V/U \to W$, for which the diagram

$$\begin{array}{ccc} V & \xrightarrow{\ f\ } & W \\ \downarrow & \nearrow_{\varphi} & \\ V/U & & \end{array}$$

is commutative.

PROOF: In any event, such a φ must satisfy $\varphi(v+U) = f(v)$, so there cannot be more than one. On the other hand, φ is well defined by $\varphi(v+U) := f(v)$, since $f(v) = f(v+a)$ for $a \in U$ follows from the linearity of f and $U \subset \operatorname{Ker} f$. Finally, the linearity of φ also follows from that of f, since $\varphi((v+U)+(v'+U)) = f(v+v') = f(v)+f(v') = \varphi(v+U)+\varphi(v'+U)$, and analogously for $\varphi(\lambda(v+U))$. □

Why does one need quotient spaces? In a first course of linear algebra they are perhaps unnecessary. But in higher mathematics, particularly in algebra and topology, quotients of all sorts occur so frequently that I thought it worthwhile to introduce you to this notion.

It would be quite understandable if you felt more at home with V and U rather than with V/U: a vector space whose vectors are subsets of another vector space? However, later on you will learn to regard U, V and U/V as equal partners ("fiber, total space, and basis") in a geometric or algebraic situation, or even experience V/U as the really useful object, for which V and U have been only preliminary raw material.

4.5 Rotations and Reflections of the Plane

A section for physicists

We consider the Euclidean vector space $\mathbb{R}^2$ and ask ourselves the following question: which linear maps $\mathbb{R}^2 \to \mathbb{R}^2$ respect the inner product, that is, for which 2×2 matrices A do we have $\langle Ax, Ay\rangle = \langle x, y\rangle$ for all $x, y \in \mathbb{R}^2$? First of all we shall attempt to answer this question by pictorial considerations. The columns are known to be the images of the unit vectors; so we consider the two unit vectors $e_1 = (1,0)$ and $e_2 = (0,1)$. The image vector Ae_1 of e_1, shown in Fig. 35, must have length 1, since $\|Ae_1\|^2 = \langle Ae_1, Ae_1\rangle = \langle e_1, e_1\rangle = 1$.

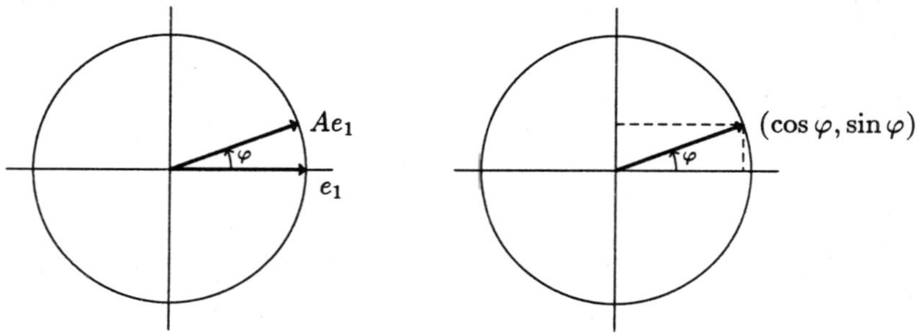

Fig. 35. The image Ae_1 of the first unit vector e_1

The angle swept out by e_1, as we pass in the mathematically positive sense (that is counterclockwise) to Ae_1, may be called φ. Then, written as a column, Ae_1 has the components $\cos\varphi$ and $\sin\varphi$, hence we already know how the first column of A must look,

$$A = \begin{pmatrix} \cos\varphi & * \\ \sin\varphi & * \end{pmatrix}.$$

What happens with e_2? Again we must have $\|Ae_2\| = 1$, and in addition $\langle Ae_2, Ae_1\rangle = \langle e_2, e_1\rangle = 0$, that is Ae_2 is perpendicular to Ae_1. There are thus only two possibilities, as Fig. 36 shows.

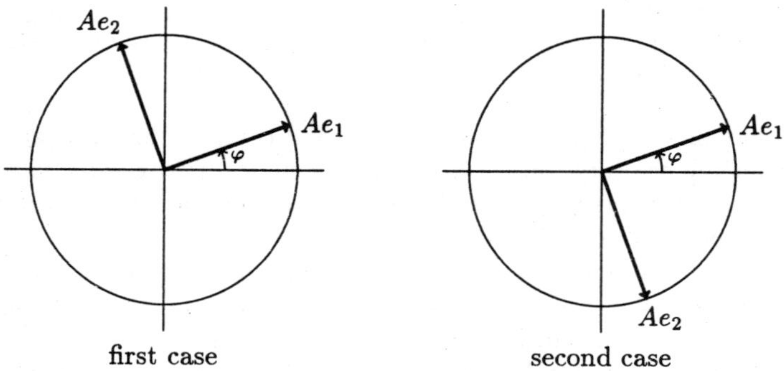

Fig. 36. The two possibilities for image Ae_2 of the second unit vector e_2

The angle formed by Ae_2 with e_1 is either $\varphi + \frac{\pi}{2}$ or $\varphi - \frac{\pi}{2}$. Since from high school mathematics we know that

$$\begin{aligned} \cos(\varphi + \tfrac{\pi}{2}) &= -\sin\varphi \\ \cos(\varphi - \tfrac{\pi}{2}) &= \sin\varphi \\ \sin(\varphi + \tfrac{\pi}{2}) &= \cos\varphi \\ \sin(\varphi - \tfrac{\pi}{2}) &= -\cos\varphi, \end{aligned}$$

the second column is either

$$\begin{pmatrix} -\sin\varphi \\ \cos\varphi \end{pmatrix} \quad \text{or} \quad \begin{pmatrix} \sin\varphi \\ -\cos\varphi \end{pmatrix},$$

and we have obtained the answer to our question.

Theorem: A 2×2 matrix A has the property

$$\langle Ax, Ay \rangle = \langle x, y \rangle$$

for all $x, y \in \mathbb{R}^2$ if and only if there exists φ so that either

$$A = \begin{pmatrix} \cos\varphi & -\sin\varphi \\ \sin\varphi & \cos\varphi \end{pmatrix} \quad \text{or} \quad A = \begin{pmatrix} \cos\varphi & \sin\varphi \\ \sin\varphi & -\cos\varphi \end{pmatrix}$$

holds.

Without pictorial arguments but assuming knowledge of sine and cosine, we proceed as follows: let $A = (a_{ij})$ and $\langle Ax, Ay \rangle = \langle x, y \rangle$ for all $x, y \in \mathbb{R}^2$. Since

$$\langle Ae_1, Ae_1 \rangle = a_{11}^2 + a_{21}^2 = 1 \quad \text{and} \quad \langle Ae_2, Ae_2 \rangle = a_{12}^2 + a_{22}^2 = 1,$$

there exist real numbers φ and ψ with $a_{11} = \cos\varphi$, $a_{21} = \sin\varphi$, $a_{22} = \cos\psi$, $a_{12} = -\sin\psi$. Then the relation $\langle Ae_1, Ae_2 \rangle = a_{11}a_{12} + a_{21}a_{22} = 0$ implies that $-\cos\varphi \sin\psi + \sin\varphi\cos\psi = \sin(\varphi - \psi) = 0$, hence that $\varphi = \psi + k\pi$ for some $k \in \mathbb{Z}$. Thus, either $\cos\varphi = \cos\psi$ and $\sin\varphi = \sin\psi$ (namely, if k is even) or $\cos\varphi = -\cos\psi$ and $\sin\varphi = -\sin\psi$ (namely, if k is odd). From this it follows that A must be of the given form. Conversely, it is an easy calculation to check that such a matrix respects the inner product, and the theorem is proved. □

Definition: The set of all 2×2 real matrices in the above theorem is denoted by $O(2)$ (from "orthogonal"); the subset

$$\{A \in O(2) \mid A = \begin{pmatrix} \cos\varphi & -\sin\varphi \\ \sin\varphi & \cos\varphi \end{pmatrix}, \quad \varphi \in \mathbb{R}\}$$

is denoted by $SO(2)$.

If one considers where a given matrix $A \in O(2)$ maps the two unit vectors e_1, e_2, and that from $x = \lambda_1 e_1 + \lambda_2 e_2$ it also follows that $Ax = \lambda_1 Ae_1 + \lambda_2 Ae_2$, it is easy to see the geometric mechanism behind A. And here there is an essential difference between the matrices from $SO(2)$ and those from $O(2) \setminus SO(2)$.

Geometrically as a map $\mathbb{R}^2 \to \mathbb{R}^2$, the matrix

$$A = \begin{pmatrix} \cos\varphi & -\sin\varphi \\ \sin\varphi & \cos\varphi \end{pmatrix} \in SO(2)$$

describes a rotation about the origin — more precisely a rotation through the angle φ in a mathematically positive sense; see Fig. 37.

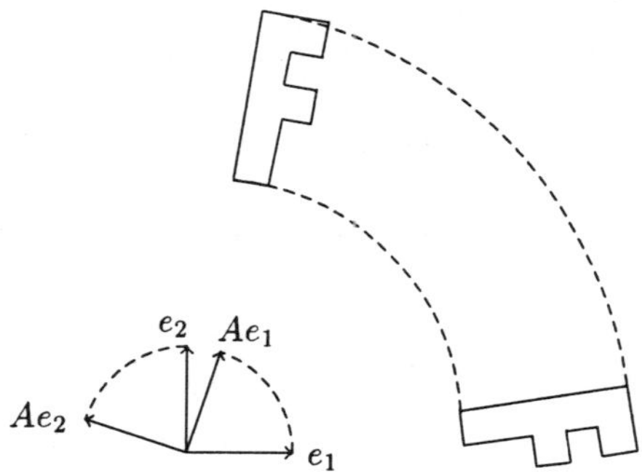

Fig. 37. Rotation through the angle φ between e_1 and Ae_1

However, the matrix

$$B = \begin{pmatrix} \cos\varphi & \sin\varphi \\ \sin\varphi & -\cos\varphi \end{pmatrix} \in O(2) \setminus SO(2)$$

describes a *reflection* in the axis making an angle $\frac{\varphi}{2}$ with $\mathbb{R} \times 0$; see Fig. 38.

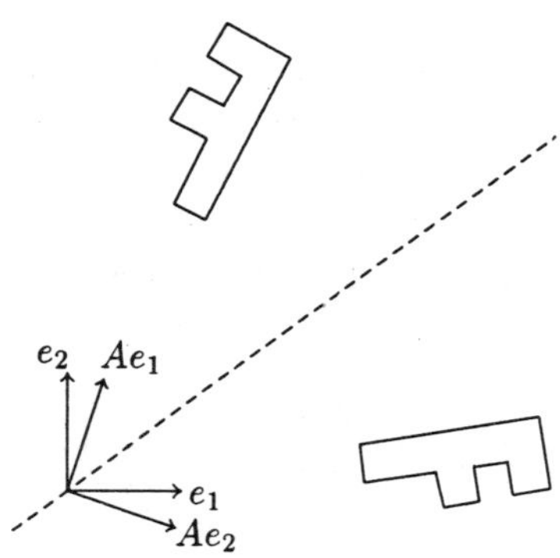

Fig. 38. Reflection in the axis bisecting the angle φ between e_1 and Ae_1

Please note that by means of

$$\begin{aligned} \mathbb{R}^2 &\longrightarrow \mathbb{R}^2 \\ \begin{pmatrix} x_1 \\ x_2 \end{pmatrix} &\longmapsto \begin{pmatrix} \cos\varphi & -\sin\varphi \\ \sin\varphi & \cos\varphi \end{pmatrix} \begin{pmatrix} x_1 \\ x_2 \end{pmatrix} \end{aligned}$$

we effect a rotation of all of $\mathbb{R}^2$ through the angle φ, not just a rotation of the "coordinate system" and the giving of the coordinates of the point x "in" the new coordinate system. This is something quite different! Let us have a look at it.

Perhaps we can best understand the "introduction of new coordinates" for a vector space if we forget for a moment the existence of the old ones. Let V be a two-dimensional real vector space. A "coordinate system" is given by a basis (v_1, v_2). One calls $L(v_1)$ and $L(v_2)$ the coordinate axes. Consider the canonical basis-isomorphism $\Phi : \mathbb{R}^2 \xrightarrow{\cong} V$, $(\lambda_1, \lambda_2) \mapsto \lambda_1 v_1 + \lambda_2 v_2$. The inverse map $\Phi^{-1} : V \xrightarrow{\cong} \mathbb{R}^2$, given by $\lambda_1 v_1 + \lambda_2 v_2 \mapsto (\lambda_1, \lambda_2)$, is then the map that assigns coordinates to each vector (see Fig. 39).

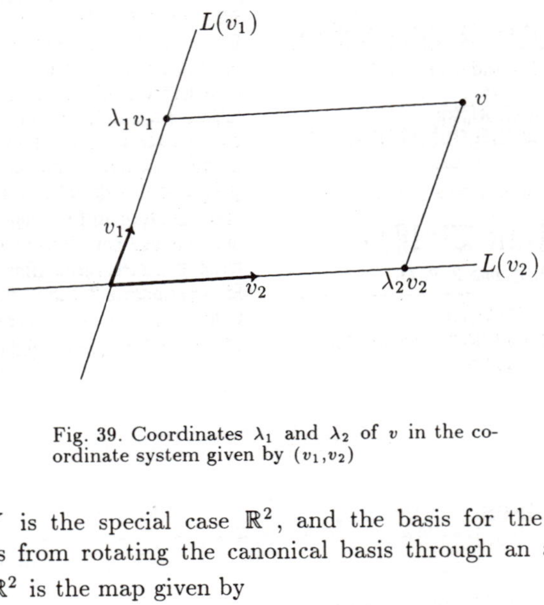

Fig. 39. Coordinates λ_1 and λ_2 of v in the coordinate system given by (v_1, v_2)

Now, if V is the special case $\mathbb{R}^2$, and the basis for the "new coordinates" results from rotating the canonical basis through an angle φ, then $\Phi : \mathbb{R}^2 \xrightarrow{\cong} \mathbb{R}^2$ is the map given by

$$\begin{pmatrix} \cos\varphi & -\sin\varphi \\ \sin\varphi & \cos\varphi \end{pmatrix}$$

(as we have already described), and if one is interested in $\Phi^{-1} : \mathbb{R}^2 \xrightarrow{\cong} \mathbb{R}^2$ because one wants to give each vector from $\mathbb{R}^2$ its "new coordinates," then this map is given by the matrix representing a rotation through $-\varphi$, that is

$$\begin{pmatrix} \cos\varphi & \sin\varphi \\ -\sin\varphi & \cos\varphi \end{pmatrix}.$$

4.6 Historical Aside

The reasons for studying mathematics are most diverse. For readers with some knowledge of German, I let Christian Wolff speak for himself, without further comment on my part.

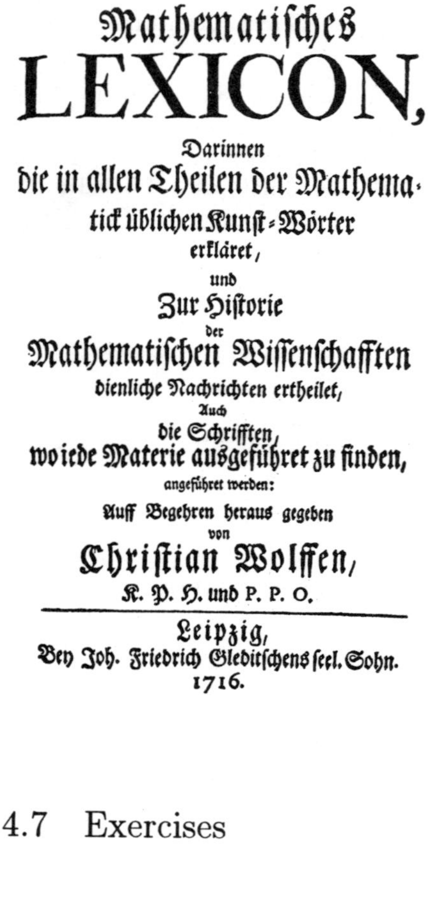

Mathematisches
LEXICON,
Darinnen
die in allen Theilen der Mathematick üblichen Kunst-Wörter
erkläret,
und
Zur Historie
der
Mathematischen Wissenschafften
dienliche Nachrichten ertheilet,
Auch
die Schrifften,
wo iede Materie ausgeführet zu finden,
angeführet werden:
Auff Begehren heraus gegeben
von
Christian Wolffen,
K. P. H. und P. P. O.

Leipzig,
Bey Joh. Friedrich Gleditschens seel. Sohn.
1716.

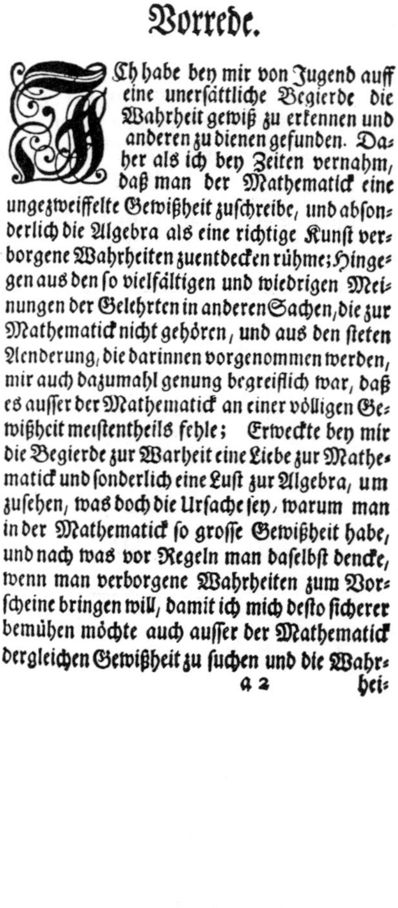

Vorrede.

Ich habe bey mir von Jugend auff eine unersättliche Begierde die Wahrheit gewiß zu erkennen und anderen zu dienen gefunden. Daher als ich bey Zeiten vernahm, daß man der Mathematick eine ungezweiffelte Gewißheit zuschreibe, und absonderlich die Algebra als eine richtige Kunst verborgene Wahrheiten zuentdecken rühme; Hingegen aus den so vielfältigen und wiedrigen Meinungen der Gelehrten in anderen Sachen, die zur Mathematick nicht gehören, und aus den steten Aenderung, die darinnen vorgenommen werden, mir auch dazumahl genung begreiflich war, daß es ausser der Mathematick an einer völligen Gewißheit meistentheils fehle; Erweckte bey mir die Begierde zur Warheit eine Liebe zur Mathematick und sonderlich eine Lust zur Algebra, um zusehen, was doch die Ursache sey, warum man in der Mathematick so grosse Gewißheit habe, und nach was vor Regeln man daselbst dencke, wenn man verborgene Wahrheiten zum Vorscheine bringen will, damit ich mich desto sicherer bemühen möchte auch ausser der Mathematick dergleichen Gewißheit zu suchen und die Wahr-

a 2 hei-

4.7 Exercises

Exercises for mathematicians

4.1: Let V and W be vector spaces over $\mathbb{F}$, let $(v_1, \ldots, v_n)$ be a basis of V, and let $f : V \to W$ be a linear map. Show that f is injective if and only if $(f(v_1), \ldots, f(v_n))$ is linearly independent.

4.2: Let $\mathbb{F}$ be a field and $\mathscr{P}_n = \{\lambda_0 + \lambda_1 t + \cdots + \lambda_n t^n \mid \lambda_i \in \mathbb{F}\}$ be the vector space of polynomials in the indeterminate t of degree $\leq n$ with coefficients in $\mathbb{F}$. If $f(t) \in \mathscr{P}_n$ and $g(t) \in \mathscr{P}_m$, the product $f(t)g(t) \in \mathscr{P}_{n+m}$ is defined in the obvious way. If it worries you that you don't really know what an "indeterminate" is, and that therefore the whole definition of $\mathscr{P}_n$ itself hangs rather in the air (and I actually expect that this does worry you), then you can simply define $\mathscr{P}_n$ as $\mathbb{F}^{n+1}$, a polynomial as $(\lambda_0, \ldots \lambda_n)$, $\lambda_i \in \mathbb{F}$, and the product by

$$(\lambda_0, \ldots \lambda_n) \cdot (\mu_0, \ldots, \mu_m) := (\sum_{i+j=0} \lambda_i \mu_j, \ldots, \sum_{i+j=n+m} \lambda_i \mu_j).$$

So we can avoid the "undetermined nature of the concept indeterminate" by means of such a simple formalization, and having seen this, we return reassured to the usual comfortable way of speaking and writing about polynomials.

We call $(1, t, \ldots, t^n)$ the canonical basis of $\mathscr{P}_n$. Determine the matrix of the linear map $\mathscr{P}_3 \to \mathscr{P}_4$, $f(t) \mapsto (2-t)f(t)$ relative to the canonical bases.

4.3: By a finite chain complex C one understands a sequence of homomorphisms

$$0 \xrightarrow{f_{n+1}} V_n \xrightarrow{f_n} \cdots \xrightarrow{f_2} V_1 \xrightarrow{f_1} V_0 \xrightarrow{f_0} 0$$

with the property that $f_i \circ f_{i+1} = 0$, i.e., $\operatorname{Im} f_{i+1} \subset \operatorname{Ker} f_i$. The quotient vector space $H_i(C) := \operatorname{Ker} f_i / \operatorname{Im} f_{i+1}$ is called the ith ***homology group*** of the complex. Show that if all the V_i are finite-dimensional, then

$$\sum_{i=0}^{n} (-1)^i \dim V_i = \sum_{i=0}^{n} (-1)^i \dim H_i(C).$$

The $*$-exercise

4*: Suppose that in the following commutative diagram of vector spaces and homomorphisms the two rows are "exact," that is, $\operatorname{Ker} f_i = \operatorname{Im} f_{i+1}$ and $\operatorname{Ker} g_i = \operatorname{Im} g_{i+1}$ for $i = 1, 2, 3$.

$$\begin{array}{ccccccccc}
V_4 & \xrightarrow{f_4} & V_3 & \xrightarrow{f_3} & V_2 & \xrightarrow{f_2} & V_1 & \xrightarrow{f_1} & V_0 \\
{\scriptstyle \text{surj.}} \downarrow {\scriptstyle \varphi_4} & & {\scriptstyle \cong} \downarrow {\scriptstyle \varphi_3} & & \downarrow {\scriptstyle \varphi_2} & & {\scriptstyle \cong} \downarrow {\scriptstyle \varphi_1} & & {\scriptstyle \text{inj.}} \downarrow {\scriptstyle \varphi_0} \\
W_4 & \xrightarrow[g_4]{} & W_3 & \xrightarrow[g_3]{} & W_2 & \xrightarrow[g_2]{} & W_1 & \xrightarrow[g_1]{} & W_0
\end{array}$$

Suppose further that the "vertical" homomorphisms in the diagram have the properties shown, thus φ_4 is surjective, φ_3 and φ_1 are isomorphisms, and φ_0 is injective. Show that under these conditions φ_2 is an isomorphism.

Exercises for physicists

4.1P: Exercise **4.1** (for mathematicians)

4.2P: Let $(V, \langle\ ,\ \rangle)$ be a Euclidean vector space and let $f : V \to V$ be a linear map. Show that $\langle f(x), f(y)\rangle = \langle x, y\rangle$ for all $x, y \in V$ if and only if $\|f(x)\| = \|x\|$ for all $x \in V$.

4.3P: Let $(V, \langle\ ,\ \rangle)$ be a two-dimensional Euclidean vector space and let $f : V \to V$ be an orthogonal linear map, i.e., $\langle f(x), f(y)\rangle = \langle x, y\rangle$ for all $x, y \in V$. We further assume that f, without being the identity on V, has a nontrivial *fixed point*, that is, a vector $v_0 \in V \setminus 0$ with $f(v_0) = v_0$. Show that relative to any orthonormal basis (e_1, e_2) of V, the matrix representing f must be an element of $O(2) \setminus SO(2)$.

CHAPTER 5

Matrix Calculus

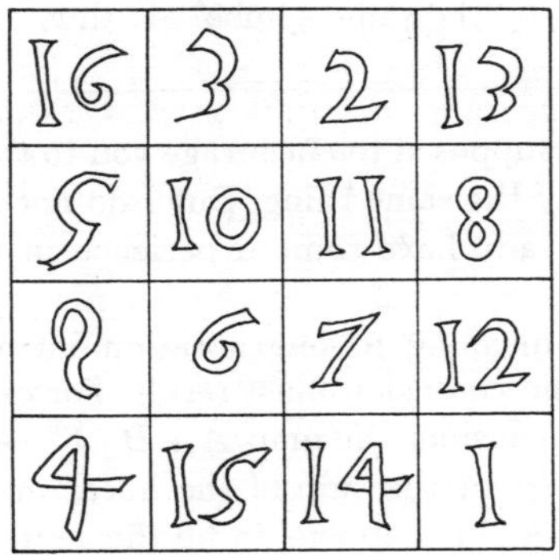

5.1 Multiplication

In this section we will talk at length on matrix multiplication, but first a word about ***addition*** and ***scalar multiplication*** in $M(m \times n, \mathbb{F})$. Instead of $A = \begin{pmatrix} a_{11} & \cdots & a_{1n} \\ \vdots & & \vdots \\ a_{m1} & \cdots & a_{mn} \end{pmatrix}$ one can also write $A = (a_{ij})_{i=1\ldots m; j=1\ldots n}$, or if it is already clear how many rows and columns A has, just simply $A = (a_{ij})$. Addition and scalar multiplication are defined element-wise, as follows.

Definition: Let (a_{ij}), $(b_{ij}) \in M(m \times n, \mathbb{F})$ and $\lambda \in \mathbb{F}$. Then

$$(a_{ij}) + (b_{ij}) := (a_{ij} + b_{ij}) \in M(m \times n, \mathbb{F}), \text{ and}$$
$$\lambda(a_{ij}) := (\lambda a_{ij}) \in M(m \times n, \mathbb{F}).$$

Fact 1: In this way $M(m \times n, \mathbb{F})$ becomes a vector space over $\mathbb{F}$. Since this vector space is only distinguished from $\mathbb{F}^{mn}$ by the manner of writing its elements (as a rectangle instead of in a long row or column), it has dimension mn.

Fact 2: The map $M(m \times n, \mathbb{F}) \to \text{Hom}(\mathbb{F}^n, \mathbb{F}^m)$, defined by associating the linear map $x \mapsto Ax$ of $\mathbb{F}^n$ into $\mathbb{F}^m$ to each matrix A, is an isomorphism of vector spaces.

Now for multiplication. Everything that we are going to say here about matrices has two aspects: conceptual and computational, depending on whether we regard matrices as linear maps $\mathbb{F}^n \to \mathbb{F}^m$ or as arrangements of numbers. We will use the same notation for both aspects, thus accepting another case of "double meaning."

Convention: For matrices $A \in M(m \times n, \mathbb{F})$ we denote the associated linear map $\mathbb{F}^n \to \mathbb{F}^m$ by the same symbol A, thus $A : \mathbb{F}^n \to \mathbb{F}^m$.

Of course, this is not supposed to encourage you to think that a matrix and a linear map are actually the same thing. But I do not need to spell out such naive warnings; you already have some experience in the domain of double meanings.

To assign a double meaning to one notation introduces certain obligations, particularly that no confusion must result. For example, if for matrices $A, B \in M(m \times n, \mathbb{F})$ we consider the map $A + B : \mathbb{F}^n \to \mathbb{F}^m$, do we mean the matrix sum as linear map, or the sum of the linear maps $A, B : \mathbb{F}^n \to \mathbb{F}^m$? Never mind: in both cases this happens to be the same map, so here there is no danger of confusion, and the same holds for λA, $\lambda \in \mathbb{F}$.

Now we will use exactly the same pattern to define matrix multiplication: the product of two matrices will coincide as linear map with the composition

$$AB : \mathbb{F}^n \xrightarrow{B} \mathbb{F}^m \xrightarrow{A} \mathbb{F}^r.$$

What does this mean for the numerical evaluation of AB? Note, first of all, that we are not going to multiply arbitrary matrices $A \in M(r \times m, \mathbb{F})$ and $B \in M(s \times n, \mathbb{F})$ together, since one can only compose

$$\mathbb{F}^n \xrightarrow{B} \mathbb{F}^s, \ \mathbb{F}^m \xrightarrow{A} \mathbb{F}^r$$

as AB, if $s = m$. Therefore the matrix product defines a map

$$\begin{aligned} M(r \times m, \mathbb{F}) \times M(m \times n, \mathbb{F}) &\longrightarrow M(r \times n, \mathbb{F}), \\ (A, B) &\longmapsto AB. \end{aligned}$$

In order to determine the formula for AB, one must simply work out the image of the jth unit vector: $e_j \longmapsto Be_j \longmapsto ABe_j$, because this is the jth column of AB:

$$\begin{pmatrix} 0 \\ \cdot \\ 1 \\ \cdot \\ 0 \end{pmatrix} \longmapsto \begin{pmatrix} b_{1j} \\ \vdots \\ b_{mj} \end{pmatrix} \longmapsto \begin{pmatrix} a_{11}b_{1j} + \cdots + a_{1m}b_{mj} \\ \vdots \\ a_{r1}b_{1j} + \cdots + a_{rm}b_{mj} \end{pmatrix}$$

(compare with Section 4.2). Therefore, $\sum_{k=1}^{m} a_{ik}b_{kj}$ is the ith element of the jth column of AB. Let us use this formula as definition of the product in the main text, and note its significance for the composition of linear maps as a consequence.

Definition: If $A = (a_{ik}) \in M(r \times m, \mathbb{F})$ and $B = (b_{kj}) \in M(m \times n, \mathbb{F})$, the ***product*** $AB \in M(r \times n, \mathbb{F})$ is defined by

$$AB := \Big(\sum_{k=1}^{m} a_{ik} b_{kj}\Big)_{\substack{i=1,\dots,r \\ j=1,\dots,n}} .$$

Fact 3: As one can easily work out, the matrix product corresponds to the composition of the associated linear maps; thus the diagram

$$\begin{array}{ccc} \mathbb{F}^n & \xrightarrow{B} & \mathbb{F}^m \\ & \searrow^{AB} & \downarrow A \\ & & \mathbb{F}^r \end{array}$$

is commutative. In particular our notation does not involve any danger of confusion in connection with the apparently different definitions of AB as matrix product and AB as composition of linear maps.

The same also holds if we describe homomorphisms between finite-dimensional vector spaces by means of matrices relative to bases in these spaces. If V, W, and Y are vector spaces and $(v_1, \dots, v_n)$, $(w_1, \dots, w_m)$, and $(y_1, \dots, y_r)$, respectively, are bases for them, then because of the commutativity of the diagram

$$\begin{array}{ccccc} V & \xrightarrow{f} & W & \xrightarrow{g} & Y \\ \cong\uparrow & & \cong\uparrow & & \cong\uparrow \\ \mathbb{F}^n & \xrightarrow{B} & \mathbb{F}^m & \xrightarrow{A} & \mathbb{F}^r \end{array}$$

in which the vertical arrows are the basis isomorphisms, and A and B are the matrices describing g and f with respect to these bases, the matrix AB corresponds to the homomorphism gf.

In connection with the explicit working out of a matrix product, note the following pattern in Fig. 40.

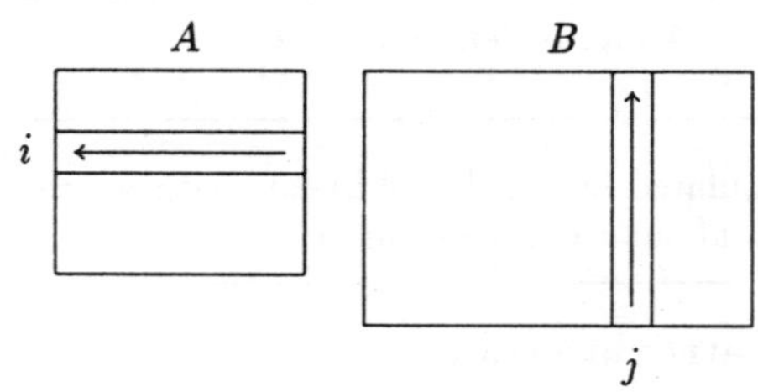

Fig. 40. Calculation of the element c_{ij} of the product matrix $C := AB$

We interpret this by saying that the element lying in the ith row and jth column of the product is calculated from the ith row of A and jth column of B by "matching together-multiplying-summing up," as has already been explained in Section 4.2 for the application of an $m \times n$ matrix to an n-tuple written as a column. Hence only the jth column of B plays a role in the jth column of AB. For example, if the jth column of B is zero, so is the jth column of AB, and similarly for the rows of AB and A.

Note also that the pattern visualizes which matrices can be multiplied: the rows of A must be as long as the columns of B, if it is to be possible to form the product AB, that is, A must have the same number of columns as B has rows.

Fact 4: Matrix multiplication is associative: $A(BC) = (AB)C$, and distributive with respect to addition: $A(B + C) = AB + AC$ and $(A + B)C = AC + BC$. This follows from the corresponding properties for linear maps.

These are properties that one expects of a multiplication. However, for matrix multiplication there exist important departures from the rules for multiplying numbers.

Remark: Multiplication of (square) matrices is neither commutative (there exist matrices with $AB \neq BA$) nor free of "zero divisors" (there exist matrices $A \neq 0$, $B \neq 0$, with $AB = 0$).

PROOF: Choosing $A = \begin{pmatrix} 0 & 1 \\ 0 & 1 \end{pmatrix}$, $B = \begin{pmatrix} 1 & 1 \\ 0 & 0 \end{pmatrix}$, we obtain examples of both

$$AB = \begin{pmatrix} 0 & 1 \\ 0 & 1 \end{pmatrix} \begin{pmatrix} 1 & 1 \\ 0 & 0 \end{pmatrix} = \begin{pmatrix} 0 & 0 \\ 0 & 0 \end{pmatrix} = 0$$

and

$$BA = \begin{pmatrix} 1 & 1 \\ 0 & 0 \end{pmatrix} \begin{pmatrix} 0 & 1 \\ 0 & 1 \end{pmatrix} = \begin{pmatrix} 0 & 2 \\ 0 & 0 \end{pmatrix} \neq AB.$$

Definition: A matrix A is called ***invertible*** if the associated linear map is an isomorphism. The matrix of the inverse map is then called the matrix ***inverse*** to A and is denoted by A^{-1}.

Using our accumulated knowledge of linear maps we can easily form a collection of assertions about the inverse matrix.

Remarks on matrix inversion:

(1) Each invertible matrix A is square, i.e., $A \in M(n \times n, \mathbb{F})$.

(2) If $A \in M(n \times n, \mathbb{F})$ is invertible, then so is A^{-1}, and $(A^{-1})^{-1} = A$.

(3) If $A, B \in M(n \times n, \mathbb{F})$ are invertible, then the product AB is also invertible and we have $(AB)^{-1} = B^{-1}A^{-1}$.

(4) If $A, B \in M(n \times n, \mathbb{F})$ and E_n (or simply E) denotes the matrix of the identity map $\mathbb{F}^n \to \mathbb{F}^n$, i.e.

$$E = \begin{pmatrix} 1 & & \\ & \ddots & \\ & & 1 \end{pmatrix},$$

then B is the matrix inverse to A if and only if $AB = BA = E$, and indeed the sharper statement (5) holds:

(5) If $A, B \in M(n \times n, \mathbb{F})$, then $AB = E \Leftrightarrow BA = E \Leftrightarrow B = A^{-1}$.

PROOFS: Invertible matrices are square, since $\mathbb{F}^n \not\cong \mathbb{F}^m$ for $m \neq n$. Assertion (4) follows from Exercise 2 of Chapter 1, and (2) and (3) are in any case clear; see

$$\mathbb{F}^n \underset{B^{-1}}{\overset{B}{\rightleftarrows}} \mathbb{F}^n \underset{A^{-1}}{\overset{A}{\rightleftarrows}} \mathbb{F}^n$$

It remains to prove (5). We know already that $B = A^{-1}$ implies the other two statements. Therefore, suppose first that $AB = E$. Then A is surjective, since for each $y \in \mathbb{F}$ we have $A(By) = Ey = y$. Now apply Fact 5 from the end of Section 4.1: this says that A must be even *bijective*. Hence A^{-1} exists, and it remains to check whether $A^{-1} = B$. It suffices to show that not only does $AB = E$ but also $BA = E$. We have

$$BA = (A^{-1}A)BA = A^{-1}(AB)A = A^{-1}EA = A^{-1}A = E,$$

and therefore we have shown that $AB = E \iff BA = E$ so that (5) follows from (4). □

Not so easy to find is a method for the explicit determination of A^{-1}. We will come back to this in Section 5.6.

5.2 The Rank of a Matrix

In Section 4.1 we defined the rank of a linear map f to be $\dim \operatorname{Im} f$. Correspondingly, one takes the rank of a matrix $A \in M(m \times n, \mathbb{F})$ to be the dimension of the image of $A : \mathbb{F}^n \to \mathbb{F}^m$. This number is also equal to the maximal length of a linearly independent r-tuple of columns of A, since the columns, as images of the unit vectors, generate $\operatorname{Im} A$, and hence, by the basis extension theorem, there exists a basis of $\operatorname{Im} A$ consisting of columns of A, and there can exist no *longer* linearly independent r-tuple of columns (why?).

Definition: If $A \in M(m \times n, \mathbb{F})$, then

$$\operatorname{rk} A := \dim \operatorname{Im}(A : \mathbb{F}^n \to \mathbb{F}^m)$$

is called the ***rank*** of the matrix A. The maximal number of linearly independent columns is called the ***column rank*** of A, and the maximal number of linearly independent rows is the ***row rank*** of A.

Fact: The rank of a matrix is equal to its column rank.

Theorem: For each matrix A, column rank and row rank are equal.

PROOF: For the purposes of this proof we shall say that a row or column is *linearly superfluous* if it can be expressed as a linear combination of the other rows or columns. If we reduce the size of a matrix by omitting a linearly superflous column it is clear that the *column* rank is unchanged. We will show first that the row rank does not change either.

Assume that the jth column is linearly superfluous in the matrix A. Then for each row or linear combination of rows, the jth entry is linearly superfluous (in the one-dimensional space $\mathbb{F}$!). This is clear: one forms the jth entry as a linear combination of the remaining elements by using the same coefficients as were used to express the jth column as a combination of the others.

From this it follows that a linear combination of rows from A is already zero if the corresponding row combination *without* the jth column is zero. Therefore, the matrix A and the matrix formed by omitting a linearly superfluous column both have the same number of linearly independent rows, that is, the same row rank. This was what we wanted to prove first.

In the same way omission of a linearly superfluous *row* does not alter the *column* rank (and of course not the row rank). Now we reduce the size of our matrix by repeated omission of linearly superfluous rows and columns until this is no longer possible. At this point we obtain a possibly much smaller matrix A', which, however, still has the same row and column ranks as A.

That A' has no linearly superfluous rows and columns implies that the rows and columns of A' are linearly independent: row rank equals the number of rows, and column rank equals the number of columns. But then A' must be square, since the length of a linearly independent r-tuple of vectors cannot exceed the dimension of the space. Hence row and column ranks are equal. □

This was a rather "pedestrian" sort of proof; Exercise 11.1 in Chapter 11 will be concerned with a more conceptual one.

5.3 Elementary Transformations

The so-called "elementary row and column transformations" are in practice perhaps the most important techniques in matrix calculus. In this chapter we need them to determine the rank; in the next chapter, for calculations with determinants; and in Chapter 7 for the solution of systems of linear equations.

Definition: There are three kinds of ***elementary row transformations*** for a matrix $A \in M(m \times n, \mathbb{F})$, namely

(R1) the interchange of two rows
(R2) multiplication of a row by a scalar $\lambda \neq 0$, $\lambda \in \mathbb{F}$, and
(R3) addition of an arbitrary multiple of one row to another row (not to the same!).

Elementary column transformations (C1), (C2), (C3) are defined analogously.

After a series of elementary transformations, a matrix may be scarcely recognizable. For example, observe how the following 3×3 matrix is "cleaned up" by transformations of type (3):

$$\begin{pmatrix} 1 & 1 & 1 \\ 2 & 2 & 2 \\ 3 & 3 & 3 \end{pmatrix} \to \begin{pmatrix} 1 & 1 & 1 \\ 0 & 0 & 0 \\ 3 & 3 & 3 \end{pmatrix} \to \begin{pmatrix} 1 & 1 & 1 \\ 0 & 0 & 0 \\ 0 & 0 & 0 \end{pmatrix} \to \begin{pmatrix} 1 & 1 & 0 \\ 0 & 0 & 0 \\ 0 & 0 & 0 \end{pmatrix} \to \begin{pmatrix} 1 & 0 & 0 \\ 0 & 0 & 0 \\ 0 & 0 & 0 \end{pmatrix}$$

In spite of these drastic changes, one important reminder of the original matrix remains — namely, the rank.

Remark 1: Elementary transformations do not alter the rank of a matrix.

It is clear that elementary row transformations do not alter the linear hull of the rows, and thus not the row rank, since this is the dimension of the linear hull. In the same way, elementary column operations do not alter the column rank. Since the row and column ranks are equal, the truth of the remark follows.

This remark leads to a marvelously simple procedure for the determination of the rank of a matrix. But first, let's introduce some terminology.

Definition: The elements a_{ii} in a matrix are called the ***diagonal elements***. The remaining elements a_{ij} are said to lie "above" (respectively "below") the diagonal depending on whether $i < j$ or $i > j$.

There are matrices for which one can simply read off the rank without further calculation. The following remark describes one type of such matrices.

Remark 2: If A is a matrix with m rows in which the first r entries on the diagonal are distinct from zero, while the last $m - r$ rows as well as all elements below the diagonal vanish, then $\operatorname{rk} A = r$.

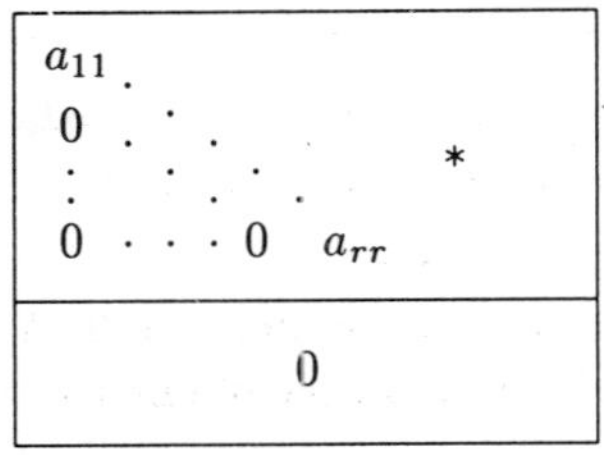

(In such a schematic representation for a matrix, the symbol $*$ means that it is irrelevant to the matter under discussion which elements occur in the region marked $*$.)

PROOF OF REMARK 2: Omitting those vanishing last $m-r$ rows obviously does not alter the rank, and the first r rows are linearly independent, because λ_1(first row)+λ_2(second row)+$\cdots$+λ_r(rth row) $=0$ implies $\lambda_1 = 0$, because $a_{11} \neq 0$, and then $\lambda_2 = 0$, because $a_{22} \neq 0$, etc. Hence the row rank is r. □

The procedure for finding the rank simply consists of using elementary transformations to bring the given matrix into the above form.

Procedure for determining the rank of a matrix:
Let $A \in M(m \times n, \mathbb{F})$ be in the form given on the left:

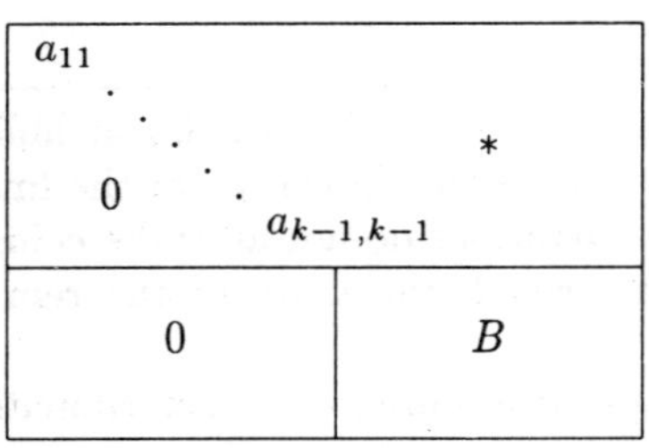

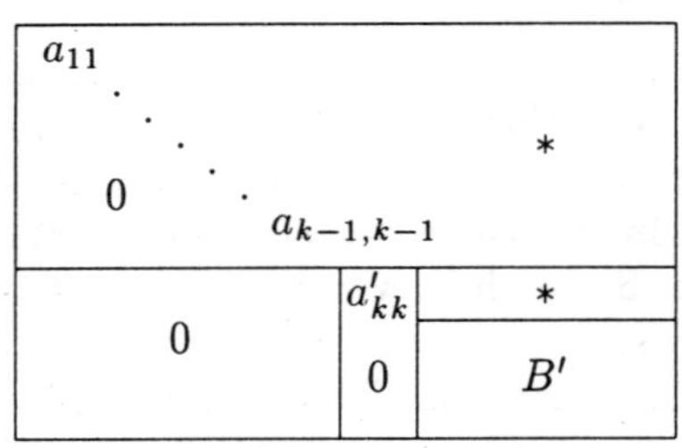

where $a_{11} \neq 0, \ldots, a_{k-1,k-1} \neq 0$ and B is an $(m-k+1) \times (n-k+1)$ matrix. If $B = 0$, then $\operatorname{rk} A = k-1$. If $B \neq 0$, there exists some $a_{ij} \neq 0$ with $i \geq k$ and $j \geq k$. If necessary, now exchange the ith and kth rows and jth and kth columns, obtaining a matrix A' with $a'_{kk} \neq 0$. By elementary transformations of type (R3) this can be brought into the form shown on the right.

If one begins this process with $k = 0$ (which means that initially the matrix does not have to satisfy any special conditions), and continues it until the residual matrix, denoted B', is either zero or has disappeared altogether, then one obtains a matrix as in Remark 2. One knows the rank of this matrix, and hence the rank of the given matrix A.

5.4 Test

(1) Let $A \in M(2 \times 3, \mathbb{F})$, $B \in M(2 \times 3, \mathbb{F})$. Then

☐ $A + B \in M(2 \times 3, \mathbb{F})$.
☐ $A + B \in M(4 \times 6, \mathbb{F})$.
☐ $A + B \in M(4 \times 9, \mathbb{F})$.

(2) For which of the following 3×3 matrices A do we have $AB = BA = B$ for all $B \in M(3 \times 3, \mathbb{F})$?

☐ $A = \begin{pmatrix} 1 & 0 & 0 \\ 0 & 1 & 0 \\ 0 & 0 & 1 \end{pmatrix}$ ☐ $A = \begin{pmatrix} 0 & 0 & 1 \\ 0 & 1 & 0 \\ 1 & 0 & 0 \end{pmatrix}$ ☐ $A = \begin{pmatrix} 1 & 1 & 1 \\ 1 & 1 & 1 \\ 1 & 1 & 1 \end{pmatrix}$

(3) For $A \in M(m \times n, \mathbb{F})$, we have

☐ A has m rows and n columns.
☐ A has n rows and m columns.
☐ The rows of A have length m and the columns of A have length n.

(4) Which of the following matrix products is zero?

☐ $\begin{pmatrix} 1 & -1 \\ 1 & -1 \end{pmatrix}\begin{pmatrix} 2 & 3 \\ -2 & -3 \end{pmatrix}$ ☐ $\begin{pmatrix} -1 & 1 \\ -1 & 1 \end{pmatrix}\begin{pmatrix} 2 & 3 \\ 2 & 3 \end{pmatrix}$ ☐ $\begin{pmatrix} 1 & -1 \\ 1 & -1 \end{pmatrix}\begin{pmatrix} 2 & 2 \\ 3 & 3 \end{pmatrix}$

(5) Which of the following properties does matrix multiplication lack?

☐ associativity ☐ commutativity ☐ distributivity

(6) For $A \in M(n \times n, \mathbb{F})$ we have:

☐ $\operatorname{rk} A = n \Rightarrow A$ is invertible, but there exist invertible matrices with $\operatorname{rk} A \neq n$.
☐ A is invertible $\Rightarrow \operatorname{rk} A = n$, but there exist matrices A with $\operatorname{rk} A = n$, which are not invertible.
☐ $\operatorname{rk} A = n \Leftrightarrow A$ is invertible.

(7) Which of the following transformations cannot be made elementarily?

☐ $\begin{pmatrix} 2 & 7 \\ 1 & 1 \end{pmatrix} \longrightarrow \begin{pmatrix} 2 & 7 \\ 3 & 8 \end{pmatrix}$

☐ $\begin{pmatrix} 1 & 1 \\ 2 & 7 \end{pmatrix} \longrightarrow \begin{pmatrix} 1 & 4 \\ 2 & 8 \end{pmatrix}$

☐ $\begin{pmatrix} 1 & 2 \\ 7 & 1 \end{pmatrix} \longrightarrow \begin{pmatrix} -11 & 2 \\ 1 & 1 \end{pmatrix}$

(8) Let $A \in M(m \times n, \mathbb{F})$, $B \in M(n \times m, \mathbb{F})$, so that we have

$$\mathbb{F}^n \xrightarrow{A} \mathbb{F}^m \xrightarrow{B} \mathbb{F}^n .$$

Let $BA = E_n$ ($=\mathrm{Id}_{\mathbb{F}^n}$ as linear map). Then

☐ $m \geq n$, A injective, B surjective.
☐ $m \leq n$, A surjective, B injective.
☐ $m = n$, A and B invertible (bijective).

(9) The rank of the real matrix

$$\begin{pmatrix} 5 & 5 & 5 \\ 5 & 5 & 5 \\ 5 & 5 & 5 \end{pmatrix}$$

is

☐ 1 ☐ 3 ☐ 5

(10) For $A \in M(m \times n, \mathbb{F})$ with $m \leq n$, we always have

☐ $\mathrm{rk}\, A \leq m$ ☐ $m \leq \mathrm{rk}\, A \leq n$ ☐ $n \leq \mathrm{rk}\, A$

5.5 How Does One Invert a Matrix?

A section for mathematicians

A good recipe for inverting matrices exists, but it is recommended that you do not learn the recipe by heart, but rather note the principle on which it rests. Then even if you find yourself with no more than a vague memory of the principle, you will have a chance of reconstructing the recipe; but if you forget one detail of the recipe, it is gone for good. Therefore, think once more about multiplying matrices. We may assume that we have $n \times n$ matrices, since these are the only ones for which inversion comes into question.

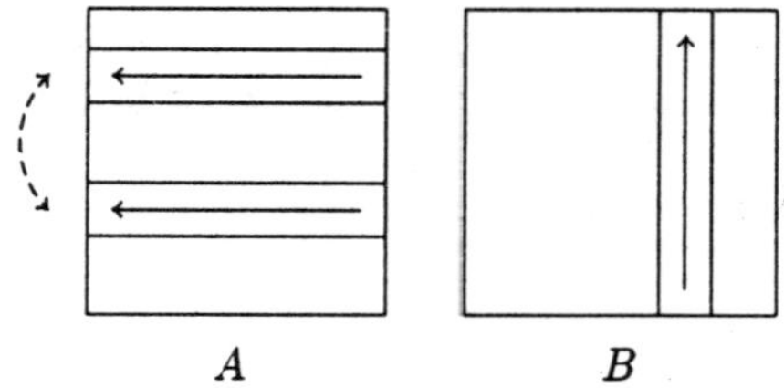

What happens with the product matrix AB, if one interchanges two rows in A (not in B!)? It is clear that the same two rows will be interchanged in the product matrix, for the ith row of the product results from the ith row of the first factor by means of the known pattern of "combining" it with the jth column of the second. In the same way multiplying the ith row of A by $\lambda \in \mathbb{F}$ has the same effect in the product, and furthermore addition of a multiple of the ith row to the jth row also carries over to the product. Hence one has the following facts.

Fact 1: If for $A, B, C \in M(n \times n, \mathbb{F})$ one has the equation $AB = C$, and one carries out the same elementary row operations on A and C obtaining A' and C', then $A'B = C'$.

Since we know that $AA^{-1} = E$, we can use this observation for our problem.

Fact 2: If E is obtained by elementary row transformations from A, the same row transformations change the matrix E into A^{-1}.

We have now to consider only how one actually turns an invertible matrix A into the identity matrix by means of row transformations. Let us remind ourselves of the types: (R1) was interchange, (R2) multiplication, and (R3) addition of a multiple.

Procedure for matrix inversion:
Let A be an $n \times n$ matrix over $\mathbb{F}$. First of all, if it is necessary, we try to make the "leading" coefficient different from zero by interchanging rows. If this is not possible, the first column is zero and A is not invertible. Suppose, therefore, that $a_{11} \neq 0$. Then multiplying the first row by $\lambda := 1/a_{11}$ replaces a_{11} by 1. Now add appropriate multiples of the first row to the other rows, so as to bring A into the form

$$\left[\begin{array}{c|c} 1 & \\ 0 & \\ \vdots & \\ 0 & \end{array}\right]$$

With this the first step is finished; in the second we try to reach the form

$$\left[\begin{array}{c|c|c} 1 & 0 & \\ 0 & 1 & \\ \vdots & 0 & \\ \vdots & \vdots & \\ 0 & 0 & \end{array}\right]$$

For this we need to interchange rows in order to have $a_{22} \neq 0$, assuming that this is not already the case, without disturbing the first row.

If this cannot be done, the second column is a multiple of the first, and the matrix is not invertible. Suppose that $a_{22} \neq 0$. By multiplying the second row by $1/a_{22}$ and by addition of suitable multiples of the second to the remaining rows, we bring the matrix into the desired form, and the second step is finished.

Either we can repeat this procedure n times, obtaining the identity matrix E, or A shows itself to be noninvertible. If A is invertible, then we obtain A^{-1} by repeating the row operations turning A into E on the rows of E in the same order.

I do not think that it is necessary to describe the kth step formally. If A is genuinely invertible, then the linear independence of the columns after $k-1$ steps guarantees the existence of an element $a_{ik} \neq 0$ with $i \geq k$, and one can proceed. Here is an illustrative example, which you should work through for yourself. There is no point in just "reading" through such a numerical example.

$$\text{Let} \quad A = \begin{pmatrix} 1 & 0 & 1 & 1 \\ 1 & 1 & 2 & 1 \\ 0 & -1 & 0 & 1 \\ 1 & 0 & 0 & 2 \end{pmatrix} \in M(4 \times 4, \mathbb{R})$$

Let's do the calculation:

$$\text{Start} \qquad A = \begin{pmatrix} 1 & 0 & 1 & 1 \\ 1 & 1 & 2 & 1 \\ 0 & -1 & 0 & 1 \\ 1 & 0 & 0 & 2 \end{pmatrix}, \begin{pmatrix} 1 & 0 & 0 & 0 \\ 0 & 1 & 0 & 0 \\ 0 & 0 & 1 & 0 \\ 0 & 0 & 0 & 1 \end{pmatrix} = E$$

$$\text{1st step} \qquad \begin{pmatrix} 1 & 0 & 1 & 1 \\ 0 & 1 & 1 & 0 \\ 0 & -1 & 0 & 1 \\ 0 & 0 & -1 & 1 \end{pmatrix}, \begin{pmatrix} 1 & 0 & 0 & 0 \\ -1 & 1 & 0 & 0 \\ 0 & 0 & 1 & 0 \\ -1 & 0 & 0 & 1 \end{pmatrix}$$

$$\text{2nd step} \qquad \begin{pmatrix} 1 & 0 & 1 & 1 \\ 0 & 1 & 1 & 0 \\ 0 & 0 & 1 & 1 \\ 0 & 0 & -1 & 1 \end{pmatrix}, \begin{pmatrix} 1 & 0 & 0 & 0 \\ -1 & 1 & 0 & 0 \\ -1 & 1 & 1 & 0 \\ -1 & 0 & 0 & 1 \end{pmatrix}$$

$$\text{3rd step} \qquad \begin{pmatrix} 1 & 0 & 0 & 0 \\ 0 & 1 & 0 & -1 \\ 0 & 0 & 1 & 1 \\ 0 & 0 & 0 & 2 \end{pmatrix}, \begin{pmatrix} 2 & -1 & -1 & 0 \\ 0 & 0 & -1 & 0 \\ -1 & 1 & 1 & 0 \\ -2 & 1 & 1 & 1 \end{pmatrix}$$

$$\text{4th step} \quad E = \begin{pmatrix} 1 & 0 & 0 & 0 \\ 0 & 1 & 0 & 0 \\ 0 & 0 & 1 & 0 \\ 0 & 0 & 0 & 1 \end{pmatrix}, \begin{pmatrix} 2 & -1 & -1 & 0 \\ -1 & \frac{1}{2} & -\frac{1}{2} & \frac{1}{2} \\ 0 & \frac{1}{2} & \frac{1}{2} & -\frac{1}{2} \\ -1 & \frac{1}{2} & \frac{1}{2} & \frac{1}{2} \end{pmatrix} = A^{-1}.$$

Result: The matrix A is invertible, and one has

$$A^{-1} = \begin{pmatrix} 2 & -1 & -1 & 0 \\ -1 & \frac{1}{2} & -\frac{1}{2} & \frac{1}{2} \\ 0 & \frac{1}{2} & \frac{1}{2} & -\frac{1}{2} \\ -1 & \frac{1}{2} & \frac{1}{2} & \frac{1}{2} \end{pmatrix}.$$

Agreed?

5.6 Rotations and Reflections (Continued)

A section for physicists

For $\varphi \in \mathbb{R}$ we introduce, just for the following discussion, the abbreviated notations

$$A_\varphi := \begin{pmatrix} \cos\varphi & -\sin\varphi \\ \sin\varphi & \cos\varphi \end{pmatrix} \in SO(2)$$

$$B_\varphi := \begin{pmatrix} \cos\varphi & \sin\varphi \\ \sin\varphi & -\cos\varphi \end{pmatrix} \in O(2) \setminus SO(2)$$

The map $A_\varphi : \mathbb{R}^2 \to \mathbb{R}^2$ is thus rotation through the angle φ, and B_φ is reflection in the axis making an angle $\varphi/2$ with $\mathbb{R} \times 0$.

How do these matrices behave with respect to multiplication, that is, what are $A_\varphi A_\psi$, $A_\varphi B_\psi$, $B_\psi A_\varphi$, and $B_\varphi B_\psi$? Before doing the calculation let us consider geometrically what must come out. If we first rotate through an angle ψ and then through an angle φ, the combined rotation is through an angle $\varphi + \psi$ (see Figs. 41a, b, and c).

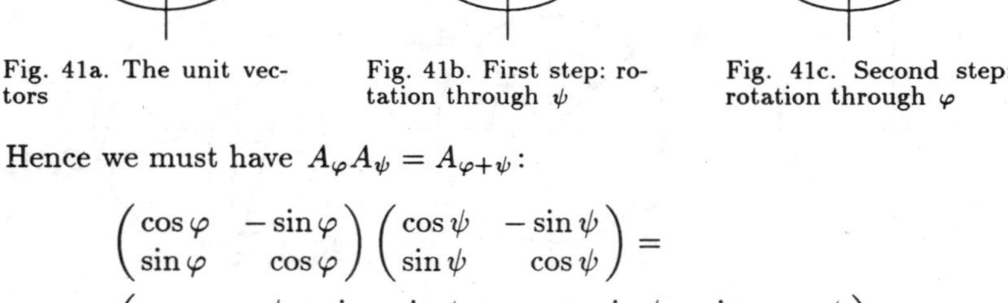

Fig. 41a. The unit vectors

Fig. 41b. First step: rotation through ψ

Fig. 41c. Second step: rotation through φ

Hence we must have $A_\varphi A_\psi = A_{\varphi+\psi}$:

$$\begin{pmatrix} \cos\varphi & -\sin\varphi \\ \sin\varphi & \cos\varphi \end{pmatrix} \begin{pmatrix} \cos\psi & -\sin\psi \\ \sin\psi & \cos\psi \end{pmatrix} =$$

$$\begin{pmatrix} \cos\varphi\cos\psi - \sin\varphi\sin\psi & -\cos\varphi\sin\psi - \sin\varphi\cos\psi \\ \sin\varphi\cos\psi + \cos\varphi\sin\psi & -\sin\varphi\sin\psi + \cos\varphi\cos\psi \end{pmatrix} =$$

$$\begin{pmatrix} \cos(\varphi+\psi) & -\sin(\varphi+\psi) \\ \sin(\varphi+\psi) & \cos(\varphi+\psi) \end{pmatrix},$$

where we have assumed, both here and in what follows, that the "addition theorem" for sine and cosine, namely,

$$\sin(\varphi + \psi) = \sin\varphi\cos\psi + \cos\varphi\sin\psi,$$
$$\cos(\varphi + \psi) = \cos\varphi\cos\psi - \sin\varphi\sin\psi$$

is known. Next consider $A_\varphi B_\psi$: we first reflect in the axis with angle $\psi/2$, and then rotate through an angle φ (see Figs. 42a, b, and c). What happens to the unit vectors (columns!)?

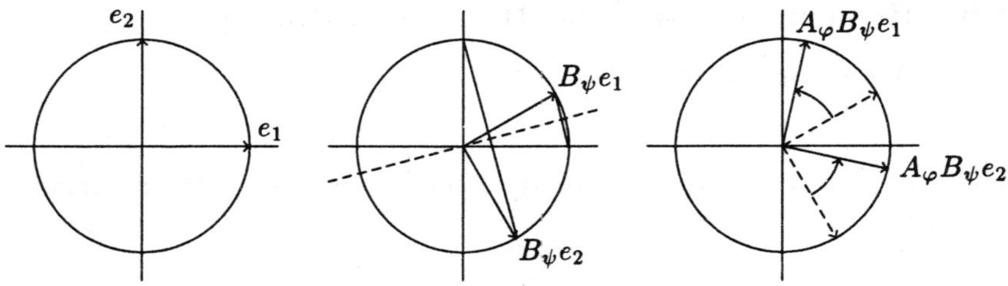

Fig. 42a. The unit vectors

Fig. 42b. First step: reflection in $\psi/2$

Fig. 42c. Second step: rotation through φ

Geometrically this gives $A_\varphi B_\psi = B_{\varphi+\psi}$. If in doubt, multiply out the matrices

$$\begin{pmatrix} \cos\varphi & -\sin\varphi \\ \sin\varphi & \cos\varphi \end{pmatrix} \begin{pmatrix} \cos\psi & \sin\psi \\ \sin\psi & -\cos\psi \end{pmatrix} =$$
$$\begin{pmatrix} \cos\varphi\cos\psi - \sin\varphi\sin\psi & \cos\varphi\sin\psi + \sin\varphi\cos\psi \\ \sin\varphi\cos\psi + \cos\varphi\sin\psi & \sin\varphi\sin\psi - \cos\varphi\cos\psi \end{pmatrix} =$$
$$\begin{pmatrix} \cos(\varphi+\psi) & \sin(\varphi+\psi) \\ \sin(\varphi+\psi) & -\cos(\varphi+\psi) \end{pmatrix} = B_{\varphi+\psi}.$$

But if we *first* rotate through an angle φ, and *then* reflect in $\psi/2$, as is shwon in Figs. 43a, b, and c,

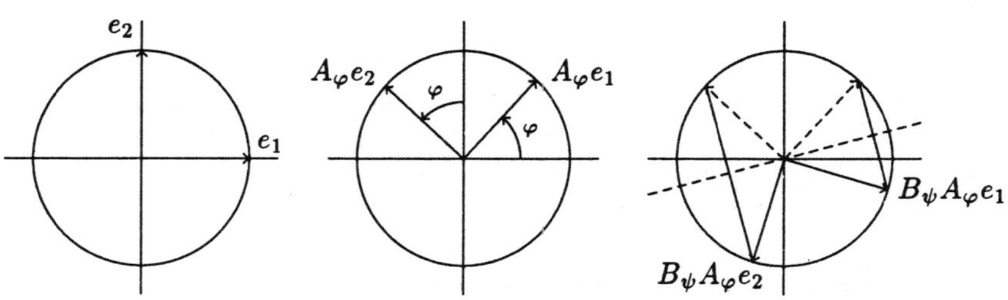

Fig. 43a. The unit vectors

Fig. 43b. First step: rotation through φ

Fig. 43c. Second step: reflection in $\psi/2$

we arrive at $B_\psi A\varphi = B_{\psi-\varphi}$. This is confirmed by

$$\begin{pmatrix} \cos\psi & \sin\psi \\ \sin\psi & -\cos\psi \end{pmatrix} \begin{pmatrix} \cos\varphi & -\sin\varphi \\ \sin\varphi & \cos\varphi \end{pmatrix} =$$

$$\begin{pmatrix} \cos\psi\cos\varphi + \sin\psi\sin\varphi & -\cos\psi\sin\varphi + \sin\psi\cos\varphi \\ \sin\psi\cos\varphi - \cos\psi\sin\varphi & -\sin\psi\sin\varphi - \cos\varphi\cos\psi \end{pmatrix} =$$

$$\begin{pmatrix} \cos(\psi - \varphi) & \sin(\psi - \varphi) \\ \sin(\psi - \varphi) & -\cos(\psi - \varphi) \end{pmatrix} = B_{\psi-\varphi}.$$

Since in general $B_{\varphi+\psi} \neq B_{\psi-\varphi}$, we have here a further example of the non-commutativity of matrix multiplication: $A_\varphi B_\psi \neq B_\varphi A_\psi$, so long as the axis making an angle of $(\psi+\varphi)/2$ with $\mathbb{R}\times 0$ is distinct from that making an angle $(\psi-\varphi)/2$.

Finally we want to see what happens when we compose two reflections with each other — what is $B_\varphi B_\psi$? Figs. 44a, b, and c illustrate this effect.

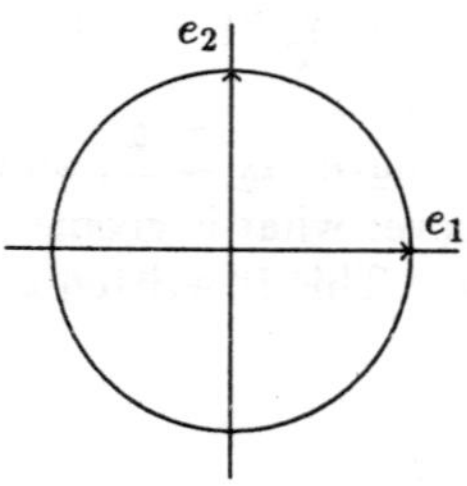

Fig. 44a. The unit vectors

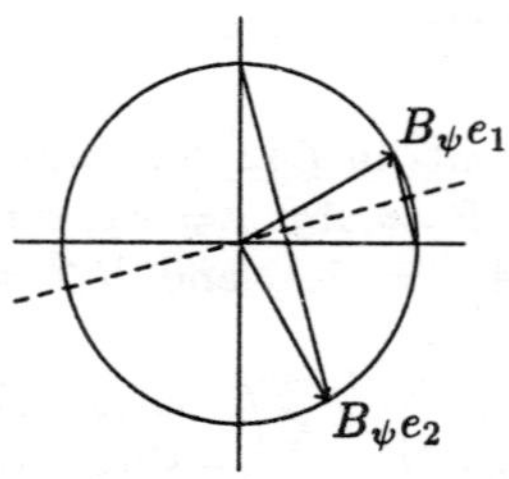

Fig. 44b. First step: reflection in $\psi/2$

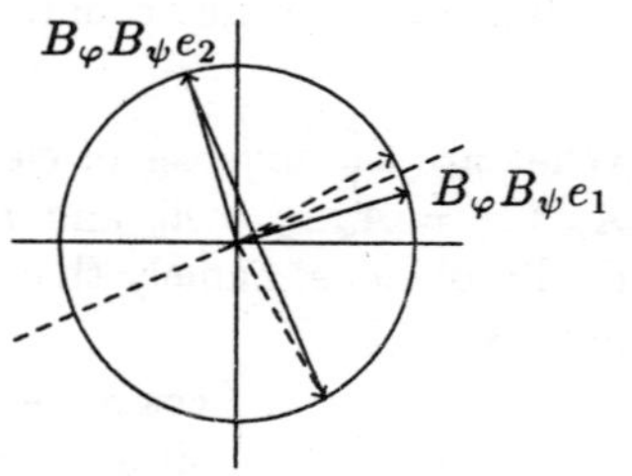

Fig. 44c. Second step: reflection in $\varphi/2$

$$\begin{pmatrix} \cos\varphi & \sin\varphi \\ \sin\varphi & -\cos\varphi \end{pmatrix} \begin{pmatrix} \cos\psi & \sin\psi \\ \sin\psi & -\cos\psi \end{pmatrix} =$$

$$\begin{pmatrix} \cos\varphi\cos\psi + \sin\varphi\sin\psi & \cos\varphi\sin\psi - \sin\varphi\cos\psi \\ \sin\varphi\cos\psi - \cos\varphi\sin\psi & \sin\varphi\sin\psi + \cos\varphi\cos\psi \end{pmatrix} =$$

$$\begin{pmatrix} \cos(\varphi - \psi) & -\sin(\varphi - \psi) \\ \sin(\varphi - \psi) & \cos(\varphi - \psi) \end{pmatrix} = A_{\varphi-\psi}.$$

From this we see that $B_\varphi B_\psi = A_{\varphi-\psi}$, and again $B_\varphi B_\psi \neq B_\varphi B_\psi$ in general.

What to remember of these formulae? I propose that one should hold on to $A_\varphi A_\psi = A_{\varphi+\psi}$, together with the following general properties of matrices in $O(2)$: rotation following rotation is rotation, rotation following reflection is reflection, reflection following rotation is reflection, and reflection following reflection is rotation.

Complete the following scheme:

Rotation through a zero angle (Identity): $A_0 = \begin{pmatrix} & \end{pmatrix}$

Rotation through 90°: $A_{\frac{\pi}{2}} = \begin{pmatrix} & \end{pmatrix}$

Rotation through 180°: $A_\pi = \begin{pmatrix} & \end{pmatrix}$

Rotation through −90° (i.e., in a clockwise direction): $A_{-\frac{\pi}{2}} = \begin{pmatrix} & \end{pmatrix}$

Reflection in $\mathbb{R} \times 0$ ("x-axis"): $B_0 = \begin{pmatrix} & \end{pmatrix}$

Reflection in the "half-angle axis" or "diagonal": $B_{\frac{\pi}{2}} = \begin{pmatrix} & \end{pmatrix}$

Reflection in the "counter diagonal": $B_{-\frac{\pi}{2}} = \begin{pmatrix} & \end{pmatrix}$

What are the inverses of elements in $O(2)$? Because we have $A_0 = E$, and $A_\varphi A_{-\varphi} = A_{\varphi-\varphi} = A_0$ and $B_\varphi B_\varphi = A_{\varphi-\varphi} = A_0$, we recover what is geometrically obvious, namely that $A_\varphi^{-1} = A_{-\varphi}$ and $B_\varphi^{-1} = B_\varphi$. This is written as follows:

$$\begin{pmatrix} \cos\varphi & -\sin\varphi \\ \sin\varphi & \cos\varphi \end{pmatrix}^{-1} = \begin{pmatrix} \cos\varphi & \sin\varphi \\ -\sin\varphi & \cos\varphi \end{pmatrix}$$

$$\begin{pmatrix} \cos\varphi & \sin\varphi \\ \sin\varphi & -\cos\varphi \end{pmatrix}^{-1} = \begin{pmatrix} \cos\varphi & \sin\varphi \\ \sin\varphi & -\cos\varphi \end{pmatrix}.$$

Thus the elements $B \in O(2) \setminus SO(2)$ all have the property that $BB = E$, or as linear maps that $B \circ B = \mathrm{Id}_{\mathbb{R}^2}$. Such maps, which when applied twice give the identity (which means that they are their own inverses), are called *involutions*. Among the elements of $SO(2)$ there are two further involutions (which?).

5.7 Historical Aside

How old would you estimate that matrix calculus is? Ten years, 100 years, 1000 years, maybe even known to the ancient Egyptians?

Matrix calculus is actually rather over one hundred years old; its originator is the English mathematician Arthur Cayley. In the year 1855 several notes by Cayley appeared in *Crelle's Journal;* in one of these the label "matrix" appears for the first time for rectangular (and in particular for square) arrays of numbers:

No. 3.

Remarques sur la notation des fonctions algébriques.

Je me sers de la notation

$$\begin{vmatrix} \alpha, & \beta, & \gamma, & \dots \\ \alpha', & \beta', & \gamma', & \dots \\ \alpha'', & \beta'', & \gamma'', & \dots \\ \dots & \dots & \dots & \dots \end{vmatrix}$$

pour représenter ce que j'appelle une *matrice;* savoir un *système* de quantités rangées en forme de *carré,* mais d'ailleurs tout à fait *indépendantes* (je ne parle pas ici des *matrices rectangulaires*). Cette notation me parait très commode pour la théorie des équations *linéaires;* j'écris par ex:

$$(\xi, \eta, \zeta \dots) = \begin{vmatrix} \alpha, & \beta, & \gamma & \dots \\ \alpha', & \beta', & \gamma' & \dots \\ \alpha'', & \beta'', & \gamma'' & \dots \\ \dots & \dots & \dots & \dots \end{vmatrix} \widehat{\ }(x, y, z \dots)$$

Three years later Cayley's foundational work on matrix calculus appeared. Of course, one had "known" about rectangular arrays of numbers for a long time — one has only to think for example of Albrecht Dürer's magic square, appearing on the etching "Melancholia" from the year 1514. But what does "know" mean here? Anyone who is familiar with numbers can write down a rectangular array of numbers. Cayley's conceptual achievement was that by introducing matrices as mathematical objects in their own right, he was the first to realize their potential for algebraic operations. The easy way, in which we define new algebraic objects ("A vector space is a triple $(V, +, \cdot)$, consisting of ... ") today, does not have a long history. Before this numbers and geometric figures were essentially the only objects of mathematics. It is with this in mind that one must see the introduction of matrices.

5.8 Exercises

Exercises for mathematicians

5.1: Show that if $A, B \in M(n \times n, \mathbb{F})$ then

$$\operatorname{rk} A + \operatorname{rk} B - n \leq \operatorname{rk} AB \leq \min(\operatorname{rk} A, \operatorname{rk} B).$$

Hint: use the dimension formula for linear transformations.

5.2: Let (v_1, v_2, v_3, v_4) be linearly independent elements of the real vector space V. If

$$\begin{aligned} w_1 &= \phantom{-v_1+{}} v_2 - v_3 + 2v_4 \\ w_2 &= v_1 + 2v_2 - v_3 - v_4 \\ w_3 &= -v_1 + v_2 + v_3 + v_4 \,, \end{aligned}$$

show that (w_1, w_2, w_3) is linearly independent.

Here a theoretical argument, as in Exercise 3.1, is no help; it comes down to the actual coefficients and one must do the calculation. Hint: first show that the linear independence of (w_1, w_2, w_3) is equivalent to a certain matrix having rank 3, and then use the procedure for determining rank to find the rank of this matrix.

5.3: For which values of λ, is the real matrix

$$A_\lambda := \begin{pmatrix} 1 & \lambda & 0 & 0 \\ \lambda & 1 & 0 & 0 \\ 0 & \lambda & 1 & 0 \\ 0 & 0 & \lambda & 1 \end{pmatrix}$$

invertible? For these values of λ determine the inverse matrix A_λ^{-1}.

The $*$-exercise

5*: Let V be a finite-dimensional vector space over $\mathbb{F}$ and $f: V \to V$ an endomorphism. Show that if with respect to all bases f is represented by the same matrix A, i.e., $A = \Phi^{-1} f \Phi$ for all isomorphisms $\Phi : \mathbb{F} \xrightarrow{\cong} V$, then there exists some $\lambda \in \mathbb{F}$ with $f = \lambda \mathrm{Id}_V$.

Exercises for physicists

5.1P: Give examples of two matrices $A, B \in M(6 \times 6, \mathbb{R})$ with the following properties: $\operatorname{rk} A = \operatorname{rk} B = 3, AB = 0$. (Giving such matrices naturally entails the proof (insofar as it is not obvious) that A and B do actually have the stated properties.)

5.2P: Exercise 5.2 (for mathematicians)

5.3P: Let

$$H_t := \begin{pmatrix} \sin 2\pi t & \sin \frac{\pi}{6} t \\ \cos 2\pi t & \cos \frac{\pi}{6} t \end{pmatrix} \quad \text{for } t \in \mathbb{R}.$$

For each t with $0 \le t < 12$, determine the rank of the matrix H_t, and in particular determine those values of t for which this rank equals 1.

CHAPTER 6

Determinants

6.1 Determinants

Each square matrix A over $\mathbb{F}$ has a *determinant,* $\det A \in \mathbb{F}$. We need the concept of the determinant in linear algebra first for certain considerations (mainly theoretical) concerning matrix inversion and the solution of systems of linear equations. Later we will again meet the determinant in discussing eigenvalues. Outside linear algebra the determinant is also important, for example, in the integration of functions of several variables, since it is closely tied to the concept of volume. But in the present chapter, we will simply consider determinants as a tool in matrix calculus, thus you need to learn what the determinant is and how to handle it.

Theorem 1 and subsequent definition: There is a unique map

$$\det : M(n \times n, \mathbb{F}) \to \mathbb{F}$$

with the following properties:

(i) det is linear in each row,

(ii) if the (row) rank is smaller than n, then $\det A = 0$,

(iii) $\det E = 1$.

This map $\det : M(n \times n, \mathbb{F}) \to \mathbb{F}$ is called the ***determinant***, and the number $\det A \in \mathbb{F}$ is called the ***determinant of*** A.

> By "linear in each row" we mean the following: if all the rows except the ith are given in some matrix scheme, each element $x \in \mathbb{F}^n$ gives an extension to a complete $n \times n$ matrix A_x: one has only to introduce x as the ith row. The map $\det : M(n \times n, \mathbb{F}) \to \mathbb{F}$ is called linear in the ith row if the map $\mathbb{F}^n \to \mathbb{F}$ given by $x \mapsto \det A_x$ is always linear.

This definition is of course no practical instruction for working out the determinant of a given matrix. If you are still of the opinion that the most important information about a mathematical object is a "formula" for "working it out," then you are certainly in the company of most educated laypeople, but as a professional mathematician you ought to throw such prejudices overboard. In most mathematical contexts where you come into contact with determinants, it is not a matter of calculating the determinant of a given matrix to two places of decimals, but rather of knowing the *properties* of the whole map $\det : M(n \times n, \mathbb{F}) \to \mathbb{F}$. (But please do not conclude from this that among mathematicians it is good taste not to be *able* to work out a determinant!)

PROOF OF THE THEOREM. (a) Proof of uniqueness: in order to check that there exists at most one map $\det : M(n \times n, \mathbb{F}) \to \mathbb{F}$ with the properties (i) to (iii), we will first prove a lemma, which also has uses outside this proof, in that it tells us how the determinant reacts to elementary row transformations.

> **Lemma:** Let $\det : M(n \times n, \mathbb{F}) \to \mathbb{F}$ be any map with properties (i) and (ii). Then the following holds:
> (1) If the matrix A' is obtained from A by interchanging two rows, then $\det A' = -\det A$.
> (2) If the matrix A' is obtained from A by multiplying some row by $\lambda \in \mathbb{F}$, then $\det A' = \lambda \det A$.
> (3) If the matrix A' is obtained from A by adding a multiple of one row to another row, then $\det A' = \det A$.

PROOF OF THE LEMMA: Assertion (2) follows immediately from the linearity of det in the rows.

For (3): Consider the matrix A'', which we obtain from A if instead of *adding* a multiple of one row to another, we *replace* this other row by the appropriate multiple. Then $\operatorname{rk} A'' < n$, and therefore $\det A'' = 0$. By linearity in the rows (here in the "other" row) it now follows that $\det A' = \det A + \det A'' = \det A$.

For (1): Let i and j label the rows to be interchanged. If we add row j to row i, by (3) we obtain from A a matrix A_1 with $\det A = \det A_1$, and similarly from A' a matrix A'_1 with $\det A' = \det A'_1$.

The matrices A_1 and A'_1 formed in this way are distinct only in the ith row: in the jth row they both have the sum of the ith and jth rows of A. Because of linearity in the ith row we then have $\det A_1 + \det A'_1 = \det B$, where B is a matrix whose ith and jth rows are both equal to the sum of the ith and jth rows of A. Therefore, $\operatorname{rk} B < n$, hence $\det B = \det A_1 + \det A'_1 = \det A + \det A' = 0$, and thus $\det A' = -\det A$. □

The lemma is now proved, and we continue in the proof of the uniqueness assertion of Theorem 1. As a consequence of the lemma we have: if det and det′ are two maps with properties (i) and (ii), and the matrix $\widetilde{A}$ is obtained from A by elementary row transformations, then $\det A = \det' A$ not only *implies*, but is actually *equivalent* to, $\det \widetilde{A} = \det' \widetilde{A}$, since any elementary row transformations can be reversed by elementary row transformations.

Let det and det′ next satisfy (i), (ii), and (iii). We want to show that $\det A = \det' A$ for all $A \in M(n \times n, \mathbb{F})$. This follows from (ii) if A is such that $\operatorname{rk} A < n$. Suppose therefore that $\operatorname{rk} A = n$. Then A can be changed by means of elementary row transformations to E; this is already known to readers of Section 5.5 on matrix inversion, and otherwise one proceeds by induction as follows. If we already have

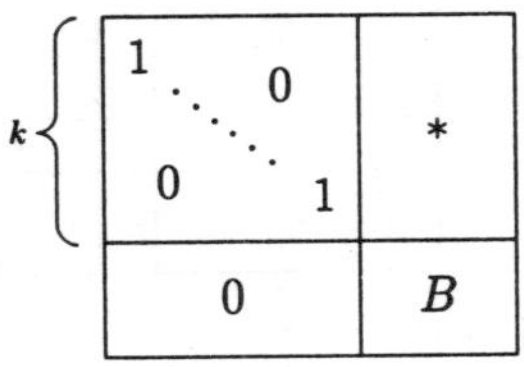

(for $k < n$), then by interchanging two of the last $n - k$ rows we can make the $(k+1, k+1)$ entry in the matrix nonzero. For if the first column of B were trivial, the $(k+1)$th column of the whole matrix would be linearly superfluous, contradicting rank $= n$. Transformations of types (R2) and (R3) then push the induction forward. Hence one can change A into E, and from $\det E = \det' E = 1$ it follows that $\det A = \det' A$. □

(b) Proof of existence: the existence of a map $\det : M(n \times n, \mathbb{F}) \to \mathbb{F}$ with properties (i) – (iii) can be proved by induction. For $n = 1$ it is clear that $M(1 \times 1, \mathbb{F}) \to \mathbb{F}$, $(a) \mapsto a$, has these properties. Now suppose that we have already defined a determinant for $(n-1) \times (n-1)$ matrices.

Definition: If $A \in M(n \times n, \mathbb{F})$, let A_{ij} denote the $(n-1) \times (n-1)$ matrix obtained from A by omission of the ith row and jth column.

With this notation and our inductive assumption we can now define a map $\det : M(n \times n, \mathbb{F}) \to \mathbb{F}$ as follows. Choose some arbitrary but then fixed j with $1 \leq j \leq n$, and set

$$\det A := \sum_{i=1}^{n} (-1)^{i+j} a_{ij} \det A_{ij}.$$

We want to show that this map $\det : M(n \times n, \mathbb{F}) \to \mathbb{F}$ has properties (i), (ii), and (iii).

Property (i): Linearity in the kth row of A follows since each summand

$$(-1)^{i+j} a_{ij} \det A_{ij}$$

has this linearity property. For $k \neq i$ this is because $\det : M(n-1\times n-1, \mathbb{F}) \to \mathbb{F}$ has the corresponding property, and a_{ij} does not depend on the kth row. For $k = i$ note that

$$\begin{aligned} M(n \times n, \mathbb{F}) &\longrightarrow \mathbb{F} \\ A &\longmapsto a_{ij} \end{aligned}$$

is linear in the ith row, and A_{ij} is independent of the ith row of A (which has been omitted!).

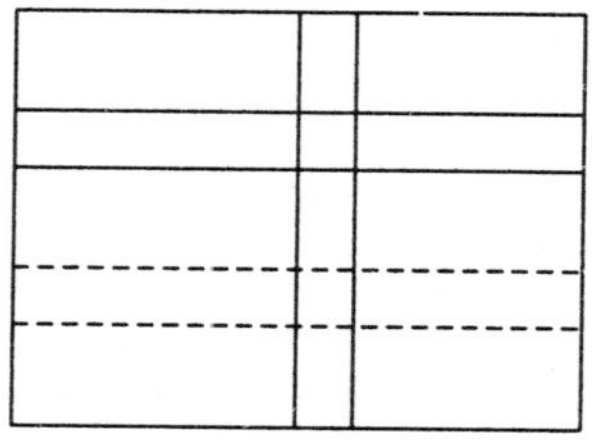

Fig. 45. Dependence of a_{ij} and A_{ij} from the kth row of A has to be considered separately for the case $k=i$ and for the case $k\neq i$.

Property (ii): Let $\operatorname{rk} A < n$. Then there exists some row that can be formed as a linear combination of the others. From this it follows that by elementary row transformations of type (R3) we can reduce this row to zero. A matrix with a zero row has determinant zero — this follows from linearity, which we have already proved. We must therefore show that elementary row transformations of type (R3) do not alter the determinant. Because we have already proved linearity, it will suffice to show that the determinant of a matrix having two equal rows vanishes. So let us assume that the rth and sth rows are the same in A. Then by the inductive assumption,

$$\sum_i (-1)^{i+j} a_{ij} \det A_{ij} = (-1)^{r+j} a_{rj} \det A_{rj} + (-1)^{s+j} a_{sj} \det A_{sj},$$

because all other summands vanish, given that the A_{ij} concerned have two equal rows. How do A_{rj} and A_{sj} differ? If r and s are adjacent, then $A_{rj} = A_{sj}$ anyway, since when two equal rows are next to each other, it is irrelevant which one we strike out. If there is exactly one other row between the rth and sth rows, thus $|r - s| = 2$, then one can change A_{rj} into A_{sj} by means of a single exchange of rows. More generally: if $|r - s| = t$, one can change A_{rj} into A_{sj} by means of $t - 1$ such moves. Since by the inductive assumption for $(n-1)\times(n-1)$ matrices interchange of rows changes the sign of the determinant, and since $a_{rj} = a_{sj}$ because of the equality of the rth and sth rows, we have

$$\begin{aligned} \det A &= (-1)^{r+j} a_{rj} \det A_{rj} + (-1)^{s+j} a_{sj} \det A_{sj} \\ &= (-1)^{r+j} a_{rj} \det A_{rj} + (-1)^{s+j} a_{rj} (-1)^{r-s+1} \det A_{rj} \\ &= ((-1)^{r+j} + (-1)^{r+j+1}) a_{rj} \det A_{rj} = 0. \end{aligned}$$

Property (iii): Here

$$E_n = \begin{pmatrix} 1 & & \\ & \ddots & \\ & & 1 \end{pmatrix} = (\delta_{ij})_{i,j=1,..,n},$$

where $\delta_{ij} = 0$ for $i \neq j$ and 1 for $i = j$ (Kronecker symbol). Therefore in the sum $\det E_n = \sum_i (-1)^{i+j}\delta_{ij} \det E_{nij}$, there is only one summand distinct from 0, and this equals $(-1)^{j+j}\delta_{jj} \det E_{njj}$. But $E_{njj} = E_{n-1}$, hence $\det E_n = \det E_{n-1} = 1$, and this completes the proof. □

6.2 Determination of Determinants

Even though the defining Theorem 1 has not given us a direct recipe for calculating determinants, in the course of the proof we have actually learned ways to do this, in particular the "expansion formula."

Column expansion formula: The formula for the determinant of an $n \times n$ matrix A obtained in the above proof,

$$\det A = \sum_{i=1}^{n} (-1)^{i+j} a_{ij} \det A_{ij},$$

is called the ***expansion of the determinant by the jth column.***

Since for the (1×1) matrix A we have $\det(a) = a$, for (2×2) matrices expansion by the first column gives

$$\det \begin{pmatrix} a & b \\ c & d \end{pmatrix} = ad - bc.$$

Once this subtraction of products formed by crossovers has become a habit, we can easily work out determinants of three-row matrices by expansion, for example by the first column:

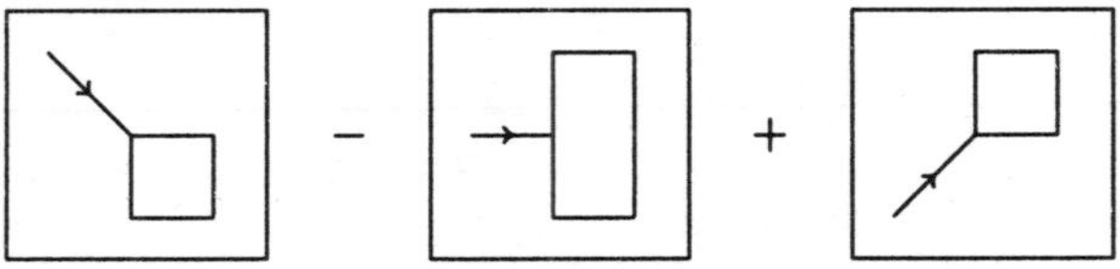

This diagram is meant to illustrate the calculation

$$\det \begin{pmatrix} a_{11} & a_{12} & a_{13} \\ a_{21} & a_{22} & a_{23} \\ a_{31} & a_{32} & a_{33} \end{pmatrix} =$$

$$a_{11} \det \begin{pmatrix} a_{22} & a_{23} \\ a_{32} & a_{33} \end{pmatrix} - a_{21} \det \begin{pmatrix} a_{12} & a_{13} \\ a_{32} & a_{33} \end{pmatrix} + a_{31} \det \begin{pmatrix} a_{12} & a_{13} \\ a_{22} & a_{23} \end{pmatrix}.$$

But already for (4×4) matrices recursive expansion of the determinant by means of an expansion formula has ceased to be economical. It still does good service in certain situations; for example, by induction and expansion by the first column we can easily prove the following facts.

Lemma: If $A \in M(n \times n, \mathbb{F})$ is an upper triangular matrix. i.e. all elements beneath the diagonal are zero, or $a_{ij} = 0$ for $i > j$,

$$\begin{pmatrix} a_{11} & & * \\ & \ddots & \\ 0 & & a_{nn} \end{pmatrix}$$

the determinant is the product of the diagonal elements:

$$\det A = a_{11} \cdot \ldots \cdot a_{nn}.$$

Corollary (Procedure for computing determinants of large matrices): In order to determine the determinant of $A \in M(n \times n, \mathbb{F})$, use elementary row transformations of types (R1) and (R3) (interchange of rows and addition of row multiples to other rows) to change A into an upper triangular matrix

$$A' = \begin{pmatrix} a'_{11} & & * \\ & \ddots & \\ 0 & & a'_{nn} \end{pmatrix},$$

which can always be done. If one has used r row interchanges,

$$\det A = (-1)^r \det A' = (-1)^r a'_{11} \cdot \ldots \cdot a'_{nn}.$$

6.3 The Determinant of the Transposed Matrix

Just as in Chapter 5 we spoke of row rank and column rank, we ought to refer to the determinant defined in Theorem 1 as the "row determinant," since property (i) refers to rows ("linear in each row"). In the same way we can define a "column determinant" by requiring the map $M(n \times n, \mathbb{F}) \to \mathbb{F}$ to be linear in each column and also to satisfy the conditions of vanishing for matrices of rank $< n$ and taking the value 1 for the identity matrix. However, we will show that the column determinant equals the row determinant, and hence that one does not need to introduce these names; one can simply speak of the determinant — similarly to the rank. One can best formulate all this by introducing the concept of the *transposed matrix*.

Definition: If $A = (a_{ij}) \in M(m \times n, \mathbb{F})$, then the matrix

$$A^t = (a^t_{ij}) \in M(n \times m, \mathbb{F})$$

with $a^t_{ij} := a_{ji}$ is called the ***transpose*** or the ***transposed matrix*** for A.

Thus one obtains A^t from A by writing the columns as rows:

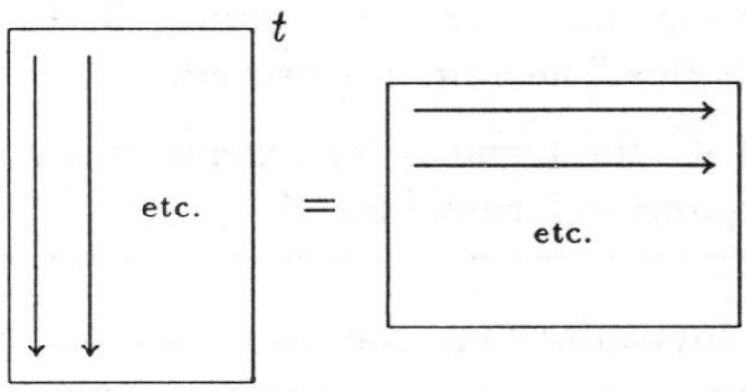

One can also describe transposition as "reflection in the diagonal," since each matrix element a_{ij} is moved from its (i, j) position to the mirror image position (j, i):

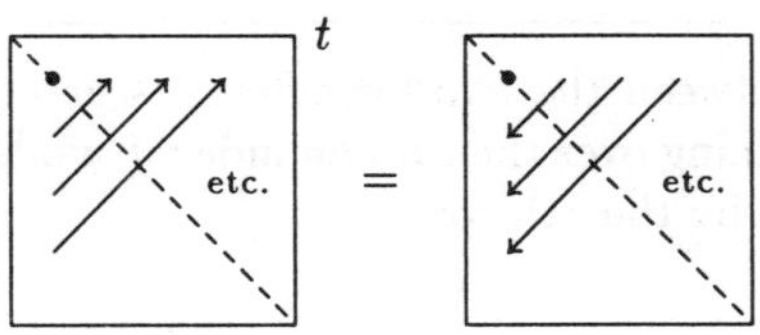

If one considers the matrix A as a linear map $A : \mathbb{F}^n \to \mathbb{F}^m$, then the transpose is a map in the opposite direction:

$$\mathbb{F}^n \xleftarrow{A^t} \mathbb{F}^m,$$

because $A^t \in M(n \times m, \mathbb{F})$. This also fits with the transpose of a matrix product $\mathbb{F}^r \xrightarrow{B} \mathbb{F}^m \xrightarrow{A} \mathbb{F}^n$ (first B, then A) being the product of the transposes in the *reverse* order, $\mathbb{F}^r \xleftarrow{B^t} \mathbb{F}^m \xleftarrow{A^t} \mathbb{F}^n$, (first A^t, then B^t).

Remark: We have $(AB)^t = B^t A^t$, since the elements c_{ij} of the product matrix $C := AB$ are by definition $c_{ij} = \sum_k a_{ik} b_{kj}$, hence $c^t_{ij} = c_{ji} = \sum_k a_{jk} b_{ki} = \sum_k b^t_{ik} a^t_{kj}$.

Furthermore it is also clear that transposition is linear, that is, $(A+B)^t = A^t + B^t$, $(\lambda A)^t = \lambda A^t$, and that $(A^t)^t = A$. We do not need to emphasize this through a seperate remark.

We can now formulate the goal of this short subsection in the language of transposed matrices.

Theorem 2: We have $\det A = \det A^t$ for all $A \in M(n \times n, \mathbb{F})$.

PROOF: Since row and column ranks are equal, and since the identity matrix E is "symmetric," that is, $E^t = E$, by Theorem 1 we have to prove only that $\det : M(n \times n, \mathbb{F}) \to \mathbb{F}$ is linear in the *columns*. However, the linearity of det in the jth column follows from the column expansion formula

$$\det A = \sum_i (-1)^{i+j} a_{ij} \det A_{ij},$$

because A_{ij} does not depend on the jth column of A, since this column has been struck out! With this Theorem 2 is proved. □

With $\det A = \det A^t$ the formula for expansion by a column, applied to A^t, gives us a row expansion formula for A.

Row expansion formula: One can also work out the determinant of an $n \times n$-matrix by means of "expansion by a row":

$$\det A = \sum_{j=1}^{n} (-1)^{i+j} a_{ij} \det A_{ij}\,.$$

The difference between the two formulae is scarcely visible to the eye. It consists in now summing over the *column* index j while holding the row index i fixed — expansion by the ith row.

6.4 Determinantal Formula for the Inverse Matrix

The row expansion formula

$$\det A = \sum_{j=1}^{n} (-1)^{i+j} a_{ij} \det A_{ij}$$

has a certain similarity with the formula for matrix multiplication. In order to put this more precisely, for each square matrix A we define the so-called adjugate matrix $\widetilde{A}$ as follows.

Definition: For $A \in M(n \times n, \mathbb{F})$ we define the matrix $\widetilde{A} \in M(n \times n, \mathbb{F})$ ***adjugate*** to A by

$$\widetilde{a}_{ij} := (-1)^{i+j} \det A_{ji}\,.$$

Thus we obtain $\widetilde{A}$ as the matrix of $(n-1)$-row *subdeterminants* $\det A_{ij}$ (remember A_{ij} results from deleting the ith row and jth column), modified by bringing in the chessboard parity $(-1)^{i+j}$ and reflection in the diagonal:

$\det A_{11}$	$-\det A_{21}$	$\det A_{31}$
$-\det A_{12}$	$\det A_{22}$	$-\det A_{32}$
$\det A_{13}$	$-\det A_{23}$	$\det A_{33}$

By the row expansion formula we then have $\det A = \sum_{j=1}^{n} a_{ij}\widetilde{a}_{ji}$, and this implies that the diagonal elements of the product matrix $A\widetilde{A}$ are all equal to $\det A$. What happens away from the diagonal? What is $\sum_{j=1}^{n} a_{ij}\widetilde{a}_{jk}$, for fixed $i \neq k$? This is the heart of the matter: the kth column

$$\begin{pmatrix} \widetilde{a}_{1k} \\ \vdots \\ \widetilde{a}_{nk} \end{pmatrix}$$

of the adjugate matrix $\widetilde{A}$ does not notice if we alter the kth row of A in any way, since $\widetilde{a}_{jk} = (-1)^{j+k} \det A_{kj}$, and A_{kj} results from deleting row k (and column j) from A. Therefore, we may just as well *replace* the kth row of A by the ith, and expand the determinant of the resulting matrix A' by its kth row to find that

$$\det A' = \sum_{j=1}^{n} a_{ij}\widetilde{a}_{jk}\,.$$

But $\det A'$ is zero, because A' has two equal rows! Thus we have proved the next theorem.

Theorem 3: If $\widetilde{A}$ is the adjugate matrix to $A \in M(n \times n, \mathbb{F})$, then

$$A\widetilde{A} = \begin{pmatrix} \det A & & \\ & \ddots & \\ & & \det A \end{pmatrix},$$

and therefore for all matrices A with nonvanishing determinant we have the inversion formula

$$A^{-1} = \frac{1}{\det A}\widetilde{A}.$$

For two- and three-row matrices this is a convenient way to work out A^{-1} explicitly. In particular, if $ad - bc \neq 0$, we have

$$\begin{pmatrix} a & b \\ c & d \end{pmatrix}^{-1} = \frac{1}{ad-bc}\begin{pmatrix} d & -b \\ -c & a \end{pmatrix}.$$

6.5 Determinants and Matrix Products

In this section we want to prove that $\det(AB) = \det A \cdot \det B$ for $n \times n$ matrices. In particular, it would follow from this that matrices of rank n have a non-vanishing determinant, because the dimension formula $\dim \operatorname{Ker} A + \operatorname{rk} A = n$ implies that such a matrix is invertible, and $\det A \neq 0$ then follows from $\det A \cdot \det A^{-1} = \det E = 1$. However, we wish to use this fact in *proving* the product formula $\det(AB) = \det A \cdot \det B$, and it is thus a good thing that we already know it: see the following lemma.

Lemma: An $n \times n$ matrix is invertible, that is, has rank n, if and only if $\det A \neq 0$.

PROOF: From the defining Theorem 1 we already know that $\operatorname{rk} A < n$ implies that $\det A = 0$. Suppose, therefore, that $\operatorname{rk} A = n$. Then one can change A into the identity matrix by means of elementary row transformations; we have seen this, for example, in the proof of the lemma in Section 6.1. The reader of Section 5.5 had seen it even earlier. Hence $\det A$ cannot be zero, because otherwise the lemma in Section 6.1 would show that $\det E = 1$ would also equal zero. □

Theorem 4: For all $A, B \in M(n \times n, \mathbb{F})$ we have

$$\det AB = \det A \cdot \det B.$$

PROOF: We want to draw Theorem 4 out of Theorem 1, almost without calculations. First fix B and consider the map

$$\begin{aligned} f : M(n \times n, \mathbb{F}) &\longrightarrow \mathbb{F} \\ A &\longmapsto \det AB. \end{aligned}$$

Then f has property (i) of the determinant: it is linear in the rows of A. For if I only change the ith row of A, then in AB this only calls for an alteration of the ith row, and with the remaining rows together with B held fixed,

$$\begin{aligned} \mathbb{F}^n &\longrightarrow \mathbb{F}^n \\ (i\text{th row of } A) &\longmapsto (i\text{th row of } AB) \end{aligned}$$

is a linear map. Hence the assertion follows from property (i) of the determinant.

As a next property of f we note that $f(A) = 0$ if $\operatorname{rk} A < n$. This is because $\operatorname{Im} AB \subset \operatorname{Im} A$, and $\operatorname{rk} A < n$ therefore implies $\operatorname{rk} AB < n$, hence f also has property (ii) from Theorem 1.

Finally, we have that $f(E) = \det EB = \det B$. Thus if $\det B \neq 0$, the map given by $A \mapsto (\det B)^{-1} \det AB$ has all three properties (i), (ii), and (iii) from Theorem 1; hence, $(\det B)^{-1} \det AB = \det A$, or $\det AB = \det A \det B$, as we wanted to prove. It only remains to consider the case where $\det B = 0$.

But then, by the previous lemma, $\operatorname{rk} B < n$, hence $\dim \operatorname{Ker} B > 0$ and therefore $\dim \operatorname{Ker} AB > 0$ as well (note $\operatorname{Ker} B \subset \operatorname{Ker} AB$). Thus $\operatorname{rk} AB < n$ and $\det AB = 0$. We conclude that $\det AB = \det A \cdot \det B$ holds in all cases. □

Corollary: If A is invertible, that is, if $\det A \neq 0$, then

$$\det A^{-1} = (\det A)^{-1},$$

since $\det A \cdot \det A^{-1} = \det AA^{-1} = \det E = 1$.

6.6 Test

(1) The determinant is a map

- □ $M(n \times n, \mathbb{F}) \to \mathbb{F}$ given by the product of the diagonal elements.
- □ $M(n \times n, \mathbb{F}) \to \mathbb{F}$, which is linear in the rows, vanishes on matrices with less than maximal rank, and takes the value 1 on E.
- □ $M(n \times n, \mathbb{F}) \to \mathbb{F}^n$, which is given by a linear combination of rows, vanishes on matrices with less than maximal rank, and takes the value 1 on E.

(2) Let $A, A' \in M(n \times n, \mathbb{F})$ and let A' be obtained from A by elementary row transformations. Which of the following statements are correct?

- □ $\det A = 0 \iff \det A' = 0$.
- □ $\det A = \det A'$.
- □ $\det A = \lambda \det A'$ for some $\lambda \in \mathbb{F}$, $\lambda \neq 0$.

(3) Which of the following assertions is correct? For $A \in M(n \times n, \mathbb{F})$ we have

- □ $\det A = 0 \Longrightarrow \operatorname{rk} A = 0$.
- □ $\det A = 0 \iff \operatorname{rk} A \leqq n - 1$.
- □ $\det A = 0 \Longrightarrow \operatorname{rk} A = n$.

(4) Which of the following statements holds for all $A, B, C \in M(n \times n, \mathbb{F})$ and all $\lambda \in \mathbb{F}$?

☐ $\det(A+B) = \det A + \det B$.
☐ $\det \lambda A = \lambda \det A$.
☐ $\det((AB)C) = \det A \det B \det C$.

(5) Which of the formulae below is called "expansion of the determinant of $A = (a_{ij})$ by the ith row"?

☐ $\det A = \sum_{i=1}^{n} (-1)^{i+j} a_{ij} \det A_{ij}$.
☐ $\det A = \sum_{j=1}^{n} (-1)^{i+j} a_{ij} \det A_{ji}$.
☐ $\det A = \sum_{j=1}^{n} (-1)^{i+j} a_{ij} \det A_{ij}$.

(6) $\det \begin{pmatrix} 1 & 0 & 1 \\ 2 & 3 & -1 \\ 0 & 1 & 1 \end{pmatrix} =$

☐ 2 ☐ 4 ☐ 6

(7) Let $E \in M(n \times n, \mathbb{F})$ be the unit matrix. Then the transposed matrix $E^t =$

☐ $\begin{pmatrix} 1 & & \\ & \ddots & \\ & & 1 \end{pmatrix}$ ☐ $\begin{pmatrix} & & 1 \\ & \cdot^{\cdot^{\cdot}} & \\ 1 & & \end{pmatrix}$ ☐ $\begin{pmatrix} t & & \\ & \ddots & \\ & & t \end{pmatrix}$

(8) $\det \begin{pmatrix} \lambda & \lambda & \lambda \\ \lambda & \lambda & \lambda \\ \lambda & \lambda & \lambda \end{pmatrix} =$

☐ 0 ☐ λ ☐ λ^3

(9) $\det \begin{pmatrix} \cos\varphi & -\sin\varphi \\ \sin\varphi & \cos\varphi \end{pmatrix} =$

☐ $\cos 2\varphi$ ☐ 0 ☐ 1

(10) Which of the following assertions is (or are) false?

☐ $\det A = 1 \Rightarrow A = E$.
☐ $\det A = 1 \Rightarrow A$ is injective as a map $\mathbb{F}^n \to \mathbb{F}^n$.
☐ $\det A = 1 \Rightarrow A$ is surjective as a map $\mathbb{F}^n \to \mathbb{F}^n$.

6.7 Determinant of an Endomorphism

It is possible to define the determinant not only for an $n \times n$ matrix, but also for an endomorphism $f : V \to V$ of an n-dimensional vector space. One takes some basis $(v_1, \ldots, v_n)$ of V, uses the associated basis isomorphism $\Phi_{(v_1,\ldots,v_n)} : \mathbb{F}^n \cong V$ in order to replace f by a matrix,

$$\begin{array}{ccc} V & \xrightarrow{f} & V \\ \Phi\uparrow\cong & & \cong\uparrow\Phi \\ \mathbb{F}^n & \xrightarrow{A} & \mathbb{F}^n, \end{array}$$

and puts $\det f := \det A$. But is $\det A$ really well defined in this fashion? Could another basis, hence another matrix, also mean another determinant for f? The following lemma removes this doubt and thus makes possible the definition of the determinant of an endomorphism.

Lemma and Definition: If $f : V \to V$ is an endomorphism of an n-dimensional vector space, and if f is represented by the matrix A relative to a basis $(v_1, \ldots, v_n)$ and by the matrix B relative to another basis $(v'_1, \ldots, v'_n)$, we have $\det A = \det B =:$ **det *f*.**

PROOF: We introduce a third matrix, C, namely the one that connects the basis isomorphisms Φ and Φ' in the diagram

$$\begin{array}{ccc} & & V \\ & \Phi' \nearrow & \uparrow \Phi \\ \mathbb{F}^n & \xrightarrow[C]{} & \mathbb{F}^n \end{array}$$

so that $C := \Phi^{-1} \circ \Phi'$. Because we have $A = \Phi^{-1} \circ f \circ \Phi$ and $B = \Phi'^{-1} \circ f \circ \Phi'$, B is the matrix product $C^{-1}AC$:

$$\begin{array}{ccccccc} & & V & \xrightarrow{f} & V & & \\ & \Phi'\nearrow & \uparrow\Phi & & \Phi\uparrow & \nwarrow\Phi' & \\ \mathbb{F}^n & \xrightarrow[C]{} & \mathbb{F}^n & \xrightarrow[A]{} & \mathbb{F}^n & \xrightarrow[C^{-1}]{} & \mathbb{F}^n \end{array}$$

Since the determinant of the product equals the product of the determinants (Theorem 4 in Section 6.5), we have

$$\det B = \det(C^{-1}) \det A \det C = \det A \det(C^{-1}C) = \det A.$$

□

For endomorphisms $A : \mathbb{F}^n \to \mathbb{F}^n$, the determinant keeps its old meaning, since we can choose $\Phi = \text{Id}$.

By considering diagrams in which endomorphisms and matrices are linked by basis isomorphisms, one can easily deduce the properties of the endomorphism determinant $\det : \text{Hom}(V, V) \to \mathbb{F}$ from those of the matrix determinant $\det : M(n \times n, \mathbb{F}) \to \mathbb{F}$. By way of example we note the following.

If V is an n-dimensional vector space over $\mathbb{F}$, the map $\text{Hom}(V, V) \to \mathbb{F}$, $f \mapsto \det f$, has among others the properties:
(1) $\det f \neq 0 \iff f$ is an isomorphism,
(2) $\det g \circ f = \det g \cdot \det f$,
(3) $\det \text{Id}_V = 1$.

The fact that it is possible to define the determinant for endomorphisms, without specifying a basis, indicates that there must be a more conceptual approach to determinants than the somewhat technical one adopted here via matrices. "Properly speaking" (which as usual means: in a certain sense) determinants belong to *multilinear algebra*, to which you will be introduced in a later course. You will then obtain a deeper understanding of the significance of the notion. In the meantime, however, let us meet another useful formula for the determinant of a matrix.

6.8 The Leibniz Formula

The Leibniz formula is

$$\det A = \sum_{\tau \in \mathcal{S}_n} \text{sign}(\tau) a_{1\tau(1)} \cdot \ldots \cdot a_{n\tau(n)},$$

and in order to understand it one must know the meanings of $\mathcal{S}_n$ and $\text{sign}(\tau)$ for $\tau \in \mathcal{S}_n$. The first is easy: $\mathcal{S}_n$ denotes the set of bijective maps

$$\tau : \{1, \ldots, n\} \xrightarrow{\cong} \{1, \ldots, n\},$$

called the *permutations* of the numbers 1 to n. As you must already know, there exist $n! := 1 \cdot 2 \cdot \ldots \cdot n$ such permutations, since choosing $\tau(1)$ from among n possibilities leaves $(n-1)$ for $\tau(2)$, etc. One needs to say rather more about the "sign" of τ, equal to ± 1.

A permutation that does no more than switch two adjacent numbers and leave the remaining $(n-2)$ fixed, is a *neighbor transposition*. Clearly, one can obtain any permutation by carrying out finitely many such transpositions, one after the other — the librarian's curse punishes those who practice this art. A permutation obtained by means of an even number of such transpositions is called even, the others are called odd, and the sign of a permutation is defined by

$$\text{sign}(\tau) := \begin{cases} +1 & \text{if } \tau \text{ even,} \\ -1 & \text{if } \tau \text{ odd.} \end{cases}$$

Then sign(Id) $= +1$, and $\text{sign}(\sigma \circ \tau) = \text{sign}(\sigma) \cdot \text{sign}(\tau)$. This follows since if σ and τ are both even or both odd, then $\sigma \circ \tau$ is even, and if only one of them, σ say, is odd, then because $\sigma = (\sigma \circ \tau) \circ \tau^{-1}$, their composition $\sigma \circ \tau$ must also be odd, and analogously for τ.

All these considerations are very simple, but something essential is lacking. Without more information it is not immediately clear that any odd permutations exist. What? Surely a transposition is an example of an odd permutation? Yes indeed, but this is not immediately obvious. We have need of a little trick.

In general we must expect a permutation τ to change the order of some pairs $i < j$, thus $\tau(j) < \tau(i)$. Denote the number of such "order reversals" by $a(\tau)$, thus

$$a(\tau) := \#\{(i,j) \mid i < j, \text{ but } \tau(j) < \tau(i)\},$$

where the symbol # stands for "number." If σ is a neighbor transposition, then

$$a(\sigma \circ \tau) = a(\tau) \pm 1,$$

since σ either creates an order reversal or undoes one. From this it follows that $a(\tau)$ is even or odd depending on whether the permutation τ is even or odd: $\text{sign}(\tau) = (-1)^{a(\tau)}$. In particular, neighbor transpositions are odd, and the same holds for more general transpositions of not necessarily adjacent elements: if r numbers lie between i and j, the transposition exchanging i and j is achieved by means of $2r+1$ neighbor transpositions.

Now we are in a position not only to read the Leibniz formula, but also to prove it.

PROOF OF THE LEIBNIZ FORMULA: We need to prove only that the map $M(n \times n, \mathbb{F}) \to \mathbb{F}$ defined by the right-hand side has properties (i), (ii), and (iii), which by Theorem 1 (in Section 6.1) characterize the determinant. Indeed, if we do this without using the concept of a determinant, we have given a further proof of the existence assertion of Theorem 1.

Property (i), linearity in the rows, is satisfied by each of the $n!$ summands and therefore by their sum. Property (iii) is also satisfied, since if A is the unit matrix, only one summand differs from zero, namely $\text{sign}(\text{Id}) \cdot \delta_{11} \cdot \ldots \cdot \delta_{nn} = 1$. Thus it remains to prove that the right-hand side of the Leibniz formula vanishes as soon as the (row) rank of A is less than n. For this it suffices, as at the analogous point in the proof of Theorem 1, to prove it for matrices A with two equal rows. Suppose that row i and row j are equal. If σ is the transposition switching i and j, and $\mathcal{A}_n$ is the set of even permutations, we can write the right-hand side of the Leibniz formula as

$$\sum_{\tau \in \mathcal{A}_n} (\text{sign}(\tau) a_{1\tau(1)} \cdot \ldots \cdot a_{n\tau(n)} + \text{sign}(\tau \circ \sigma) a_{1\tau\sigma(1)} \cdot \ldots \cdot a_{n\tau\sigma(n)})$$

But the summand $a_{1\tau(\sigma(1))} \cdot \ldots \cdot a_{n\tau(\sigma(n))}$ results from $a_{1\tau(1)} \cdot \ldots \cdot a_{n\tau(n)}$ by replacing the ith factor $a_{i\tau(i)}$ by $a_{i\tau(j)}$ and $a_{j\tau(j)}$ by $a_{j\tau(i)}$. Because of the supposed equality of the rows ($a_{ik} = a_{jk}$ for all k) this alters nothing, and the conclusion follows from $\text{sign}(\tau \circ \sigma) = -\text{sign}(\tau)$. □

6.9 Historical Aside

How unobvious it once must have been to consider matrices as mathematical objects in their own right is shown by the fact that the theory of determinants is indeed much older than that of matrices themselves. Determinants were first defined by Leibniz in connection with the solvability of systems of linear equations — namely in a letter to l'Hospital dated 28 April 1693. The term "determinant" was introduced by Gauss in his "Disquisitiones Arithmeticae" (1801).

6.10 Exercises

Exercises for mathematicians

6.1: If $A \in M(n \times n, \mathbb{F})$, any matrix obtained by the possible deletion of some rows and columns is called a ***submatrix*** of A. Show that the maximum number of rows, which occurs in a square submatrix with nonvanishing determinant, equals the rank of A.

Hint: Take another look at the proof of the theorem "row rank = column rank" in Section 5.2, and think of the relation between rank and determinant (lemma in Section 6.5).

6.2: Calculate the determinant of the $n \times n$ matrix

$$\begin{pmatrix} & & 1 \\ & \cdot^{\cdot^{\cdot}} & \\ 1 & & \end{pmatrix}.$$

The expansion formula for determinants, say by the first column, gives the induction step.

6.3: Let $M(n \times n, \mathbb{Z})$ denote the set of $n \times n$ matrices with integral coefficients. Let $A \in M(n \times n, \mathbb{Z})$. Show that there exists $B \in M(n \times n, \mathbb{Z})$ with $AB = E$ if and only if $\det A = \pm 1$.

The $*$-exercise

6*: Two bases $(v_1, \ldots, v_n)$ and $(v'_1, \ldots, v'_n)$ of the real vector space V are said to be *compatibly oriented* if the automorphism $f : V \to V$ defined by $v_i \mapsto v'_i$ has positive determinant. The set of all bases compatibly oriented with a fixed base $(v_1, \ldots, v_n)$ is called an *orientation* of V. Each n-dimensional real vector space V with $1 \leq n < \infty$ has precisely two orientations. (It has proved useful to give the zero-dimensional vector space $\{0\}$ two orientations, too, by calling the numbers ± 1 "orientations" for $\{0\}$. But this has nothing to do with the present exercise.)

By choosing one of the two orientations, one is "orienting" the vector space. More formally: an oriented vector space consists of a pair (V, or), with V an n-dimensional real vector space and "or" one of its two orientations. The bases belonging to the chosen orientation are then said to be positively oriented.

The present exercise concerns the impossibility of so orienting all k-dimensional subspaces, that no sudden "switch" in orientation occurs. Prove the following:

Let $1 \leq k < n$ and V be an n-dimensional real vector space, without loss of generality $V = \mathbb{R}^n$. Then it is impossible simultaneously to orient all k-dimensional subspaces $U \subset V$ so that each continuous map

$$(v_1, \ldots, v_k) : [0,1] \to V \times \cdots \times V,$$

which associates a linearly independent k-tuple $(v_1(t), \ldots, v_k(t))$ to each $t \in [0,1]$, and which starts positively oriented, also stays positively oriented, i.e. that $(v_1(t), \ldots, v_k(t))$ is a positively oriented basis of its linear hull for all t, provided that this is true for $t = 0$.

Exercises for physicists

6.1P: Exercise 6.1 for mathematicians.

6.2P: Exercise 6.2 for mathematicians.

6.3P: This exercise refers back to Section 3.5. Now that we are acquainted with determinants, deduce property $(5'')$ in Section 3.5 from definition (5), and property (7) from $(5'')$.

CHAPTER 7

Systems of Linear Equations

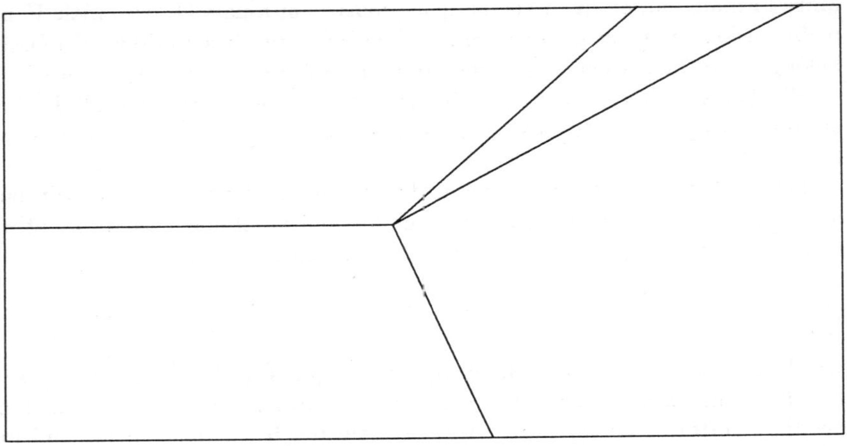

7.1 Systems of Linear Equations

Let $A = (a_{ij}) \in M(m \times n, \mathbb{F})$ and $b = (b_1, \ldots, b_m) \in \mathbb{F}^m$. Then

$$\begin{array}{ccc} a_{11}x_1 + \cdots + a_{1n}x_n & = & b_1 \\ \vdots & \vdots & \vdots \\ a_{m1}x_1 + \cdots + a_{mn}x_n & = & b_m \end{array} .$$

is called a ***system of linear equations*** for $(x_1, \ldots, x_n)$ with coefficients in $\mathbb{F}$. The $x_1, \ldots, x_n$ are the ***unknowns*** of the system. If the b_i are all zero, the system is said to be ***homogeneous***.

As in the previous sections we consider the matrix $A \in M(m \times n, \mathbb{F})$ as a linear map $A : \mathbb{F}^n \to \mathbb{F}^m$, $x \mapsto Ax$. The system of equations then reads as $Ax = b$.

Definition: The ***solution set*** of the system of equations associated to (A, b) is defined to be

$$\mathrm{Sol}(A, b) := \{x \in \mathbb{F}^n \mid Ax = b\} .$$

Instead of $\mathrm{Sol}(A,b)$ we could have written $A^{-1}(\{b\})$ or, as is usual for preimages of singleton sets, $A^{-1}(b)$. Of course the system of equations is said to be *solvable* if the solution set is nonempty.

In the present section we will collect those remarks and observations about systems of linear equations, which follow straight away from our accumulated knowledge of linear maps and matrices.

Remark 1: $Ax = b$ is solvable if and only if

$$\mathrm{rk}\begin{pmatrix} a_{11} & \cdots & a_{1n} \\ \vdots & & \vdots \\ a_{m1} & \cdots & a_{mn} \end{pmatrix} = \mathrm{rk}\begin{pmatrix} a_{11} & \cdots & a_{1n} & b_1 \\ \vdots & & \vdots & \vdots \\ a_{m1} & \cdots & a_{mn} & b_m \end{pmatrix}$$

PROOF: The columns of A generate the image of A, so if $b \in \mathrm{Im}\,A$, that is if $Ax = b$ is solvable, b can be expressed as a linear combination of the columns. Therefore the column rank of A does not change if one adjoins b as the $(n+1)$th column. Conversely, the columns of the augmented matrix (A,b) generate a subspace $V \subset \mathbb{F}^m$ containing both $\mathrm{Im}\,A$ and the vector b. From the rank condition $\dim V = \dim \mathrm{Im}\,A$ it follows that $\mathrm{Im}\,A = V$, hence that $b \in \mathrm{Im}\,A$. □

Remark 2: If $x_0 \in \mathbb{F}^n$ is a solution, that is, $Ax_0 = b$, then

$$\mathrm{Sol}(A,b) = x_0 + \mathrm{Ker}\,A := \{x_0 + x \mid x \in \mathrm{Ker}\,A\}.$$

PROOF: If $x \in \mathrm{Ker}\,A$, then $A(x_0+x) = Ax_0 = b$, hence $x_0+x \in \mathrm{Sol}(A,b)$. Conversely, if $v \in \mathrm{Sol}(A,b)$, then $A(v - x_0) = Av - Ax_0 = b - b = 0$, hence $v - x_0 \in \mathrm{Ker}\,A$, and therefore $v = x_0 + x$ for some $x \in \mathrm{Ker}\,A$ □

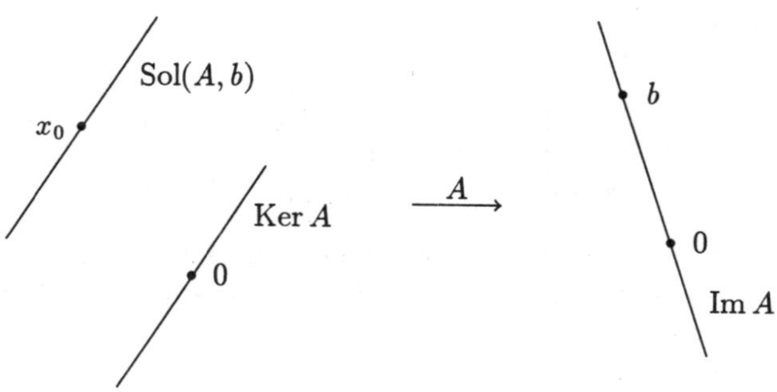

Fig. 46. $Ax=b$ is only solvable if $b \in \mathrm{Im}\,A$ (Remark 1) and then solution set is a translate of $\mathrm{Ker}\,A$ (Remark 2).

Fact 1: If x_0 is a solution of $Ax = b$ and $(v_1, \ldots, v_r)$ is a basis of $\operatorname{Ker} A$, then

$$\operatorname{Sol}(A, b) = \{x_0 + \lambda_1 v_1 + \ldots + \lambda_r v_r \mid \lambda_i \in \mathbb{F}\}.$$

Here $r = \dim \operatorname{Ker} A = n - \operatorname{rk} A$.

The relation $\dim \operatorname{Ker} A = n - \operatorname{rk} A$ follows from the dimension formula for linear maps. Remark 2 also implies the following statement.

Fact 2: A solvable system of equations $Ax = b$ is uniquely solvable if and only if $\operatorname{Ker} A = 0$, that is, $\operatorname{rk} A = n$.

For the most important case, namely $n = m$ (square matrix), we therefore have the statement below.

Fact 3: If A is square ("n equations in n unknowns"), the system of equations is uniquely solvable if and only if $\det A \neq 0$.

In this case, therefore, solvability does not depend on b at all: for *each* b there exists precisely one solution. But this is clear: $A : \mathbb{F}^n \to \mathbb{F}^n$ is bijective and the solution is nothing other than $x = A^{-1}b$. We now want to look at this "generic case," namely $\det A \neq 0$, more closely.

7.2. Cramer's Rule

If $\det A \neq 0$, there exists an explicit determinantal formula for the solution x of the system $Ax = b$, obtained as follows. Since Ax is the linear combination of the columns of A with coefficients equal to the unkowns $x_1, \ldots, x_n$, we can write $Ax = b$ in the form

$$x_1 \begin{pmatrix} a_{11} \\ \vdots \\ a_{n1} \end{pmatrix} + \cdots + x_n \begin{pmatrix} a_{1n} \\ \vdots \\ a_{nn} \end{pmatrix} = \begin{pmatrix} b_1 \\ \vdots \\ b_n \end{pmatrix}.$$

We want to discover a formula for the unknown x_i and for this exploit a little trick. Bring the column b over to the other side and subtract it from the ith summand. This gives

$$x_1 \begin{pmatrix} a_{11} \\ \vdots \\ a_{n1} \end{pmatrix} + \cdots + \begin{pmatrix} x_i a_{1i} - b_1 \\ \vdots \\ x_i a_{ni} - b_n \end{pmatrix} + \cdots + x_n \begin{pmatrix} a_{1n} \\ \vdots \\ a_{nn} \end{pmatrix} = 0,$$

so the columns of the matrix

$$\begin{pmatrix} a_{11} & \cdots & (x_i a_{1i} - b_1) & \cdots & a_{1n} \\ \vdots & & \vdots & & \vdots \\ a_{n1} & \cdots & (x_i a_{ni} - b_n) & \cdots & a_{nn} \end{pmatrix}$$

are linearly dependent, and its determinant vanishes. Because of the linearity of the determinant in the ith column, we now have that

$$x_i \det \begin{pmatrix} a_{11} & \cdots & a_{1n} \\ \vdots & & \vdots \\ a_{n1} & \cdots & a_{nn} \end{pmatrix} - \det \begin{pmatrix} a_{11} & \cdots & b_1 & \cdots & a_{1n} \\ \vdots & & \vdots & & \vdots \\ a_{n1} & \cdots & b_n & \cdots & a_{nn} \end{pmatrix} = 0,$$

where, as indicated, the determinant of the second matrix is obtained from A by replacing the ith column by b. Therefore we have the following theorem.

Theorem: If $\det A \neq 0$ and $Ax = b$, then

$$x_i = \frac{\det \begin{pmatrix} a_{11} & \cdots & b_1 & \cdots & a_{1n} \\ \vdots & & \vdots & & \vdots \\ a_{n1} & \cdots & b_n & \cdots & a_{nn} \end{pmatrix}}{\det \begin{pmatrix} a_{11} & \cdots\cdots\cdots & a_{1n} \\ \vdots & & \vdots \\ a_{n1} & \cdots\cdots\cdots & a_{nn} \end{pmatrix}}$$

for $i = 1, \ldots, n$.

This is ***Cramer's rule***, a peculiarly impractical method for solving systems of linear equations. However, Cramer's rule is of great mathematical interest, since it shows *how* the solution changes as one alters the "data" A and b of the system. With the help of the expansion formula or the Leibniz formula for the determinant, we can, for example, use Cramer's rule to show that small changes in (A, b) only lead to small changes in x. This can be made precise with ε etc. and is an important result. Put another way: Cramer's rule may be not very practical for solving a particular given system, but for studying $(A, b) \mapsto x$ as a map

$$\{A \in M(n \times n, \mathbb{F}) \mid \det A \neq 0\} \times \mathbb{F}^n \longrightarrow \mathbb{F}^n$$

it is very useful indeed.

7.3. Gaussian Elimination

Let us now turn to *the* procedure for the practical solution of systems of linear equations, namely Gaussian elimination. If we alter such a system by either interchanging two equations, multiplying an equation by $\lambda \neq 0$, or by adding a multiple of one equation to another, we do not change the solution set: clearly the solutions of the old system are also the solutions of the new, and since each of the steps can be reversed, the converse also holds. This observation lies at the root of Gaussian elimination.

Fact: If we change the augmented matrix

$$\left[\begin{array}{c|c} A & b \end{array}\right]$$

by means of elementary row transformations into a matrix

$$\left[\begin{array}{c|c} A' & b' \end{array}\right],$$

then $\text{Sol}(A, b) = \text{Sol}(A', b')$.

In contrast, elementary column transformations *do* introduce some changes to the solution set. For example, if one interchanges the first two columns, then one recovers the solution set for the old system from that of the new by interchanging the first two components in each solution n-tuple $(x_1, \ldots, x_n)$. It would indeed be possible to use column transformations as an aid to simplification, but then (other than with row transformations) we must keep track of these transformations, since finally we would have to retransform the calculated solutions of the simplified system into the solutions of the original system.

Gaussian Elimination for the Solution of Systems of Linear Equations. Let $A \in M(n \times n, \mathbb{F})$, $b \in \mathbb{F}^n$, and $\det A \neq 0$. Start with the augmented matrix

$$\left[\begin{array}{c|c} A & b \end{array}\right]$$

and if necessary use an interchange of rows to make the $(1, 1)$ or upper left matrix entry different from zero. Then add suitable multiples of the first row to the others, in order to kill the elements of the first column lying below the diagonal. This concludes the first step.

After the kth step ($k < n-1$), the $(k+1)$th proceeds as follows: by a suitable row interchange among the last $n-k$ rows, if necessary, arrange for the $(k+1, k+1)$ matrix entry to be nonzero. By addition of suitable multiples of row $(k+1)$ to those underneath it, kill all elements in column $(k+1)$ lying below the diagonal. This concludes the $(k+1)$th step.

After the $(n-1)$th step the matrix will be of the form

$$\left[\begin{array}{cccc|c} a'_{11} & \cdots & \cdots & a'_{1n} & b'_1 \\ & \ddots & & \vdots & \vdots \\ & & \ddots & \vdots & \vdots \\ 0 & & & a'_{nn} & b'_n \end{array}\right]$$

with $a'_{ii} \neq 0$ for $i = 1, \ldots, n$. We obtain the solution of $Ax = b$ by first putting

$$x_n = \frac{b'_n}{a'_{nn}}$$

and then solving recursively for the remaining unknowns:

$$\begin{aligned} x_{n-1} &= \frac{1}{a'_{n-1,n-1}}(b'_{n-1} - a'_{n-1,n}x_n), \\ x_{n-2} &= \frac{1}{a'_{n-2,n-2}}(b'_{n-2} - a'_{n-2,n}\,x_n - a'_{n-2,n-1}\,x_{n-1}), \end{aligned}$$

and so forth.

As an example, let us solve the system of equations

$$\begin{aligned} -x_1 + 2x_2 + x_3 &= -2, \\ 3x_1 - 8x_2 - 2x_3 &= 4, \\ x_1 \phantom{{}+2x_2{}} + 4x_3 &= -2. \end{aligned}$$

We start Gaussian elimination by writing down the augmented matrix:

	−1	2	1	−2
	3	−8	−2	4
	1	0	4	−2
	−1	2	1	−2
1st step	0	−2	1	−2
	0	2	5	−4
	−1	2	1	−2
2nd step	0	−2	1	−2
	0	0	6	−6

Result:
$$x_3 = \frac{1}{6}(-6) = -1,$$
$$x_2 = -\frac{1}{2}(-2+1) = \frac{1}{2},$$
$$x_1 = -(-2+1-2\cdot\frac{1}{2}) = 2.$$

The solution therefore is $x = (2, \frac{1}{2}, -1)$.

Where in this process do we actually make use of the assumption that $\det A \neq 0$? Well, if $\det A = 0$, we either fail to be able to carry out one of the steps, because it is not possible to make the diagonal element under consideration nonzero, or the last diagonal element a'_{nn} vanishes. How one can still use Gaussian elimination to solve *arbitrary* systems of linear equations is a matter to be discussed in Section 7.5 after the test.

7.4 Test

(1) A system of linear equations with coefficients in $\mathbb{F}$ is a system of equations of the following kind:

☐ $$\begin{matrix} a_{11}x_1 + \cdots + a_{1n}x_1 = b_1 \\ \vdots \qquad\quad \vdots \qquad \vdots \\ a_{n1}x_n + \cdots + a_{nn}x_n = b_n \end{matrix} \quad \text{with } a_{ij} \in \mathbb{F},\ b_i \in \mathbb{F}$$

☐ $$\begin{matrix} a_{11}x_{11} + \cdots + a_{1n}x_{1n} = b_1 \\ \vdots \qquad\quad \vdots \qquad \vdots \\ a_{n1}x_{n1} + \cdots + a_{nn}x_{nn} = b_n \end{matrix} \quad \text{with } a_{ij} \in \mathbb{F},\ b_i \in \mathbb{F}$$

☐ $$\begin{matrix} a_{11}x_1 + \cdots + a_{1n}x_n = b_1 \\ \vdots \qquad\quad \vdots \qquad \vdots \\ a_{n1}x_1 + \cdots + a_{nn}x_n = b_n \end{matrix} \quad \text{with } a_{ij} \in \mathbb{F},\ b_i \in \mathbb{F}$$

(2) If one abbreviates a system of linear equations as $Ax = b$, then

☐ $A \in M(m \times n, \mathbb{F})$, $b \in \mathbb{F}^n$.
☐ $A \in M(m \times n, \mathbb{F})$, $b \in \mathbb{F}^m$.
☐ $A \in M(m \times n, \mathbb{F})$, $b \in \mathbb{F}^n$ or $b \in \mathbb{F}^m$ (not fixed).

(3) A system of linear equations $Ax = b$ is called solvable if

☐ $Ax = b$ for all $x \in \mathbb{F}^n$.
☐ $Ax = b$ for precisely one $x \in \mathbb{F}^n$.
☐ $Ax = b$ for at least one $x \in \mathbb{F}^n$.

(4) If b is one of the columns of A, then $Ax = b$ is

☐ solvable in all cases
☐ unsolvable in all cases
☐ sometimes solvable, sometimes unsolvable, depending on A and b

(5) Let $Ax = b$ be a system of equations with square matrix A (n equations in n unknowns). Then $Ax = b$ is

☐ uniquely solvable
☐ solvable or unsolvable, depending on A, b
☐ solvable, but perhaps not uniquely, depending on A, b

(6) Suppose once more that $A \in M(n \times n, \mathbb{F})$, that is, A is square. Which of the following conditions is (or are) equivalent to the unique solvability of $Ax = b$:

☐ $\dim \operatorname{Ker} A = 0$
☐ $\dim \operatorname{Ker} A = n$
☐ $\operatorname{rk} A = n$

(7) Let $A \in M(n \times n, \mathbb{F})$ and $\det A = 0$. Then $Ax = b$ is

☐ solvable only for $b = 0$
☐ solvable for all b, possibly nonuniquely
☐ solvable only for some b, and then never uniquely

(8) Here is a more subtle question. Remember $\dim \operatorname{Ker} A + \operatorname{rk} A = n$ for $n \times n$ matrices? Good. Now let A be an $n \times n$ matrix and let $Ax = b$ have two *linearly independent* solutions. Then

☐ $\operatorname{rk} A \leq n$, and the case $\operatorname{rk} A = n$ can occur.
☐ $\operatorname{rk} A \leq n - 1$, and the case $\operatorname{rk} A = n - 1$ can occur.
☐ $\operatorname{rk} A \leq n - 2$, and the case $\operatorname{rk} A = n - 2$ can occur.

(9) Let A be a square matrix with no initial assumptions on $\det A$. If in the process of Gaussian elimination in order to obtain a solution to the system of equations $Ax = b$ the first step fails, this implies that

☐ $A = 0$.
☐ the first row of A is zero.
☐ the first column of A is zero.

(10) Let A be an $n \times n$ matrix with $\det A \neq 0$. What is the significance of being able to carry out Gaussian elimination without once having to interchange rows?

☐ A is an upper triangular matrix (nothing below the diagonal).
☐ $a_{ii} \neq 0$ for $i = 1, \ldots, n$.
☐ The *principal minor determinants*

$$\det((a_{ij})_{i,j=1,\ldots,r})$$

are nonzero for $r = 1, \ldots, n$.

7.5 More on Systems of Linear Equations

Once again we want to consider a system of equations $Ax = b$, but this time we do *not* assume that $\det A \neq 0$, and A does not even need to be square. Let $A \in M(m \times n, \mathbb{F})$ and $b \in \mathbb{F}^m$. In order to determine $\mathrm{Sol}(A, b)$ explicitly, we proceed in four steps (A)–(D) as follows.

(A) Start as for Gaussian elimination, that is, proceed as if A were square and $\det A \neq 0$. Continue as long as possible. Then after t steps the augmented matrix has the form

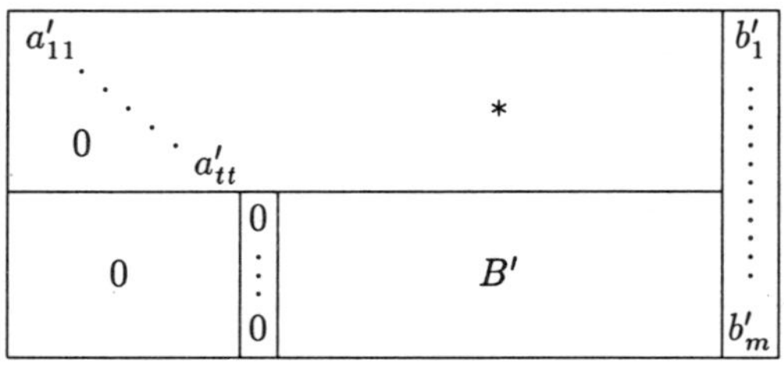

where the first t diagonal elements are nonzero, but no interchange among the last $m-t$ rows can fill the $(t+1, t+1)$ place with a nonzero element. In short, perform Gaussian elimination until it stops.

(B) Now we try and refloat the stranded Gaussian elimination with the help of *column* interchanges from among the last $n-t$ columns of A. (But doing this means interchanging some of the unknowns — make sure to keep track of the original positions of the variables!)

So long as the matrix B' does not vanish completely, we can continue the Gaussian process, finally reaching a matrix of the form

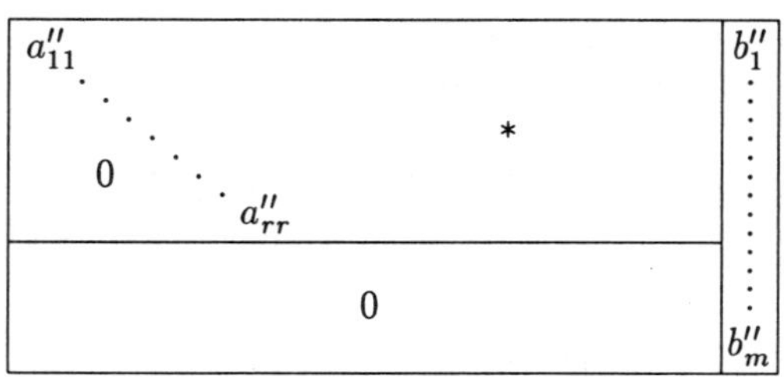

with $a''_{11} \neq 0, \ldots, a''_{rr} \neq 0$. Of course we may have $m-r=0$ or $n-r=0$.

(C) At this point one can see whether or not there exists a solution: yes if $b''_{r+1} = \cdots = b''_m = 0$, otherwise no. Let's assume the system *is* solvable. Then we can simply discard the last $m-r$ equations; this has no effect on the solutions set. There remains a system of equations with an augmented matrix of the form

$$r\left\{\left(\begin{array}{c|c|c} \underbrace{\quad T \quad}_{r} & \underbrace{\quad S \quad}_{k} & b'' \end{array}\right)\right.$$

where T (for "triangular") is an invertible upper triangular matrix

$$T = \begin{pmatrix} a''_{11} & & * \\ & \ddots & \\ 0 & & a''_{rr} \end{pmatrix}$$

and $S = (s_{ij})_{i=1\dots r, j=1\dots k}$ is just some $r \times k$ matrix. We write the n unknowns of this system as

$$y_1, \dots, y_r, z_1, \dots, z_k$$

in order to remind ourselves that these are not our original unknowns $x_1, \dots, x_n$, but, due to the column exchanges during the procedure, only a permutation of them.

The system of equations now reads $Ty + Sz = b''$ and is easy to solve. We get a particular solution $w_0 \in \mathbb{F}^n$ of it, together with a basis $(w_1, \dots, w_k)$ of the kernel of (T, S) (that is a basis of the solution space of the *homogeneous system* $Ty + Sz = 0$) by starting from the pattern

$$w_0 = \begin{pmatrix} y_1 \\ \vdots \\ y_r \\ \hline 0 \\ \vdots \\ 0 \end{pmatrix} \Bigg\}k \qquad w_1, \dots, w_k = \left(\begin{array}{cccc} y_{11} & y_{12} & \dots & y_{1k} \\ \vdots & \vdots & & \vdots \\ y_{r1} & y_{r2} & & y_{rk} \\ \hline 1 & & & \\ & 1 & & \\ & & \ddots & \\ & & & 1 \end{array}\right)$$

and determining w_0 from

$$\begin{pmatrix} a''_{11} & & * \\ & \ddots & \\ 0 & & a''_{rr} \end{pmatrix} \begin{pmatrix} y_1 \\ \vdots \\ y_r \end{pmatrix} = \begin{pmatrix} b''_1 \\ \vdots \\ b''_r \end{pmatrix}$$

and the jth basis vector w_j from

$$\begin{pmatrix} a''_{11} & & * \\ & \ddots & \\ 0 & & a''_{rr} \end{pmatrix} \begin{pmatrix} y_{1j} \\ \vdots \\ y_{rj} \end{pmatrix} = \begin{pmatrix} -s_{1j} \\ \vdots \\ -s_{rj} \end{pmatrix}$$

for $j = 1, \ldots, k$. Here, as with Gaussian elimination for an invertible matrix, we start from the bottom, thus $y_r := b''_r / a''_{rr}$, etc. The $w_1, \ldots, w_k$ genuinely belong to the kernel of $(T, S) : \mathbb{F}^n \to \mathbb{F}^r$; by construction they are linearly independent, and since by the dimension formula the dimension of the kernel is $n - r = k$, they indeed form a basis.

(D) Finally we transform the vectors

$$w_0, w_1, \ldots, w_k \in \mathbb{F}^n$$

into vectors

$$v_0, v_1, \ldots, v_k \in \mathbb{F}^n,$$

by restoring in each vector the original order of the components. That is what we needed the bookkeeping on the column interchanges for. Then

$$\text{Sol}(A, b) = \{v_0 + \lambda_1 v_1 + \cdots + \lambda_k v_k \mid \lambda_i \in \mathbb{F}\}$$

is the solution set of our original system of equations.

It is of course possible to solve the system *without* recourse to column interchanges. Then the unknowns, which we had labeled as $y_1, \ldots, y_r$ and $z_1, \ldots, z_k$, remain mixed together in the original sequence $x_1, \ldots, x_n$. Instead of arriving at the form (T, S), the matrix of the system will then evolve into the slightly more complicated *row-step* form.

7.6 Captured on Camera!

A section for physicists

Suppose that we have a billiard table, covered not with green baize but with graph paper. Suppose further that two coordinate axes are marked as shown below:

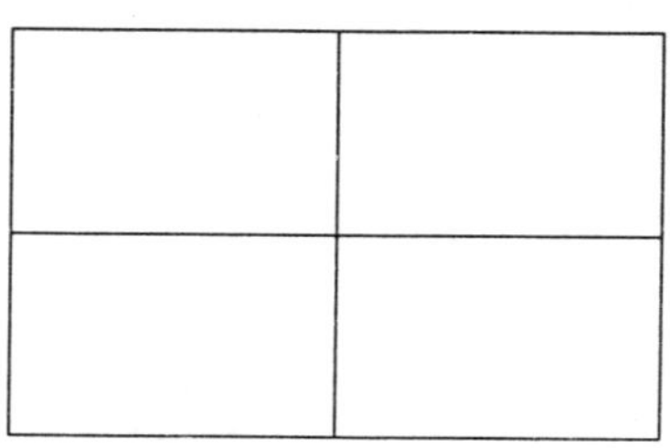

A camera mounted above the table is positioned to view the whole surface in sharp focus.

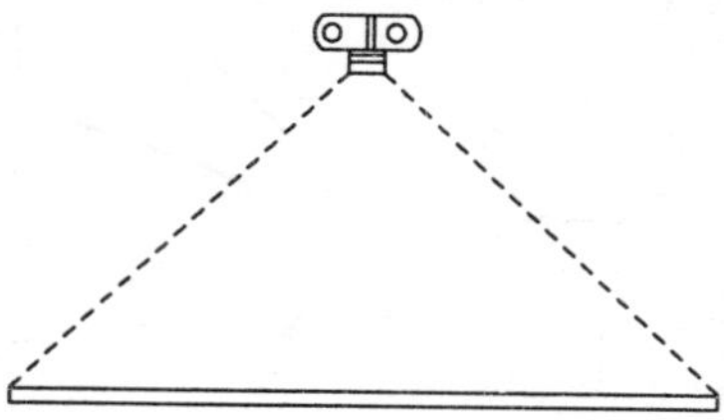

The camera has a fixed shutter speed t_0, which may be approximately equal to one second, but we do not know its value precisely. It matters only that it does not alter from picture to picture, but always equals t_0. Suppose that we are now given a number of billiard balls $K_0, K_1, K_2, \ldots$, which can be distinguished from each other by color or in some other way. Let the balls have masses $M_0, M_1, \ldots$, and suppose that the mass M_0 is known. The problem is to use the given data to determine the masses of the other balls.

Thus we want to use collision experiments and the "conservation of linear momentum" law (Berkeley Physics Course, Chapter 6) to obtain information about the involved masses. We use the camera to determine the speed and direction of the balls before and after impact. Evaluation of the measured data then leads to systems of linear equations, whose solvability we already know something about for "physical reasons." It is intriguing to compare these physical reasons with their mathematical counterparts, and at each moment to ask oneself whether it is possible to separate the physical from the purely mathematical argument.

Let us consider a simple case: the determination of M_1 and M_2 from two snapshots. If the balls K_0, K_1, and K_2 move free of any exterior forces with speeds v_0, v_1, v_2, then collide and after the collision move with speeds w_0, w_1, w_2, then conservation of momentum tells us that $M_0v_0+M_1v_1+M_2v_2 = M_0w_0 + M_1w_1 + M_2w_2$. If in particular K_1 and K_2 were at rest before the collision, then

$$M_1w_1 + M_2w_2 = M_0(v_0 - w_0). \qquad (*)$$

With the tools at our disposal we cannot measure v_i and w_i, but we can measure the distances travelled by the balls in time t_0. We proceed as follows: we position K_1 and K_2 somewhere on the table, say at the origin of the coordinate system. Then we set K_0 rolling toward K_1, and take the first picture. After the collision we take the second.

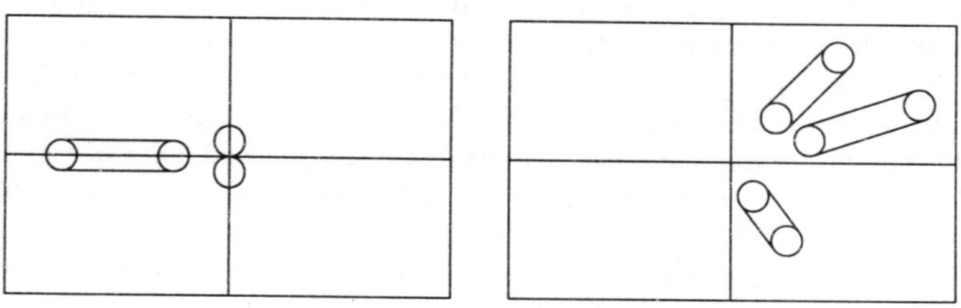

Superimposed and schematized, the images appear as follows:

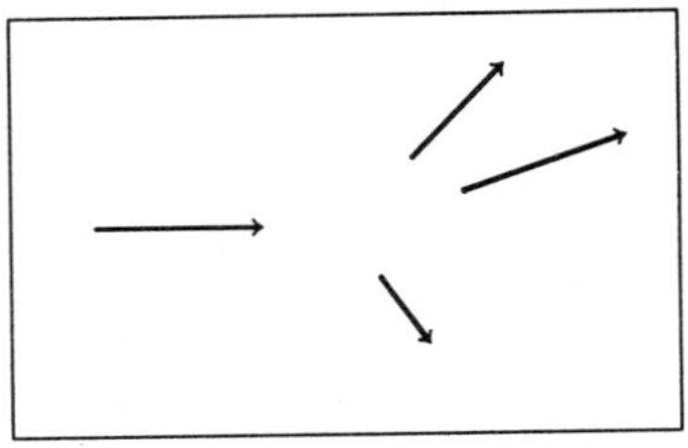

Now read off the vectors $v_0 t_0, w_1 t_0, w_2 t_0$, and $w_0 t_0 \in \mathbb{R}^2[cm]$. Conservation of momentum gives (multiply $(*)$ by t_0)

$$M_1 w_1 t_0 + M_2 w_2 t_0 = M_0(v_0 t_0 - w_0 t_0).$$

Introduce the following notation for the data and measured values:

$$\begin{aligned} M_i &= x_i \text{ gm} \\ w_i t_0 &= (a_{1i} \text{ cm},\ a_{2i} \text{ cm}) \\ M_0(v_0 t_0 - w_0 t_0) &= (b_1 \text{ gm} \cdot \text{cm},\ b_2 \text{ gm} \cdot \text{cm}). \end{aligned}$$

Then the x_i, a_{ij}, and b_i are real numbers, and we have

$$\begin{aligned} a_{11}x_1 + a_{12}x_2 &= b_1 \\ a_{21}x_1 + a_{22}x_2 &= b_2, \end{aligned}$$

a system of linear equations for unknowns x_1 and x_2. From the physical point of view we know beforehand that the system of equations must be solvable, and indeed must have a solution for which $x_1 > 0$ and $x_2 > 0$. Observe that here we are dealing with a genuinely physical argument (applicability of the law of conservation of momentum in the present situation); mathematically we can only decide on the solvability of the system of equations when we know that

$$\operatorname{rk}\begin{pmatrix} a_{11} & a_{12} \\ a_{21} & a_{22} \end{pmatrix} = \operatorname{rk}\begin{pmatrix} a_{11} & a_{12} & b_1 \\ a_{21} & a_{22} & b_2 \end{pmatrix}.$$

However we can still say mathematically that the system is *uniquely* solvable if and only if the vectors $w_1 t_0$ and $w_2 t_0$ (the columns of the coefficient matrix) are linearly independent, that is do not point in the same or directly opposite directions. The collision must therefore occur in such a way that K_1 and K_2 move in different directions, otherwise we cannot uniquely determine M_1 and M_2. In particular, none of the balls must remain motionless, for otherwise one column would be zero, $\operatorname{rk} A < 2$, and the system of equations not uniquely solvable — all for mathematical reasons. From the physical point of view it is also clear that one learns nothing about the mass of a ball if it does not move from its initial position and thus is twice photographed there. But reflect on why this clearly physically correct statement does not provide a logically sound proof for the nonunique solvability of the system of equations in such a case.

In conclusion, let me pose the following question: what is the smallest number of collision experiments that one needs in order to set up a system of linear equations to uniquely determine the masses of $K_1, \ldots, K_n$ (assuming the right kind of collisions)?

It is clear that $\frac{n}{2}$ collision experiments suffice, if n is even, and $\frac{n+1}{2}$ if n is odd: one needs only to determine the masses for pairs of balls. But if one is allowed to perform collision experiments with several balls at the same time, can one make do with a smaller number?

7.7 Historical Aside

Reference [7] in the bibliography refers to a German translation of a Chinese book from the first century B.C. with the title "Nine Books of Arithmetic Technique." And in Book VIII, "Rectangular Table" (!) appears none other than Gaussian elimination for the solution of systems of linear equations. The only difference is that the Chinese, used to writing from top to bottom, also write the rows of the matrix vertically, and as a result work with elementary column rather than with elementary row operations in order to bring the coefficient matrix into triangular form.

7.8 Remarks on the Literature

With the help of Gaussian elimination you will be able to solve numerically any given system of linear equations — in principle, that is, just as in principle anyone who knows the relation between keys and notes can play the piano. However, in reality there are difficult problems associated with the numerical solution of large systems of linear equations arising in the applications. There is a large literature in this area and new research works are continually appearing.

In order to obtain a first impression of numerical linear algebra, take a look into any of the standard works found on the library shelves. Don't worry — you are not being asked to work through them! Forget for a moment that you are a beginner, and just flip through the pages. It will become clear to you that in order to understand numerical methods, it is essential to master the basic theoretical results in a first-year linear algebra course. But you will also realize that it takes much more to become a successful numerical professional, and you will look forward to your future courses in numerical mathematics.

7.9 Exercises

Exercises for mathematicians

7.1: By determining ranks, decide whether the real system of equations below is solvable, and if so work out the solution set:

$$\begin{aligned} x_1 + 2x_2 + 3x_3 &= 1 \\ 4x_1 + 5x_2 + 6x_3 &= 2 \\ 7x_1 + 8x_2 + 9x_3 &= 3 \\ 5x_1 + 7x_2 + 9x_3 &= 4 \end{aligned}$$

7.2: Apply Gaussian elimination to the following real system of equations, decide if it is solvable, and if so determine its solution set:

$$\begin{array}{ccccccccc} x_1 & - & x_2 & + & 2x_3 & - & 3x_4 & = & 7 \\ 4x_1 & & & + & 3x_3 & + & x_4 & = & 9 \\ 2x_1 & - & 5x_2 & + & x_3 & & & = & -2 \\ 3x_1 & - & x_2 & - & x_3 & + & 2x_4 & = & -2 \end{array}$$

7.3: Prove the following:

Theorem: If $U \subset \mathbb{F}^n$ is a subspace and $x \in \mathbb{F}^n$, there exists a system of equations with coefficients in $\mathbb{F}$, having n equations and n unknowns, whose solution set equals $x + U$.

The $*$-exercise

7*: We say that two fields, $\mathbb{F}, \mathbb{F}'$, are isomorphic (written $\mathbb{F} \cong \mathbb{F}'$), if there exists a "field isomorphism" $f : \mathbb{F} \to \mathbb{F}'$, i.e., a bijective map with $f(x+y) = f(x) + f(y)$ and $f(xy) = f(x)f(y)$ for all $x, y \in \mathbb{F}$. Show that if a system of linear equations with coefficients in $\mathbb{F}$ has exactly three solutions, then $\mathbb{F} \cong \mathbb{F}_3$.

Exercises for physicists

7.1P: Let

$$\begin{aligned} a_{11}x_1 + \cdots + a_{1n}x_n &= b_1 \\ \vdots \qquad\qquad \vdots \quad &\quad \vdots \\ a_{n1}x_1 + \cdots + a_{nn}x_n &= b_n \end{aligned}$$

be a system of linear equations with real coefficients. Let $\langle\ ,\ \rangle : \mathbb{R}^n \times \mathbb{R}^n \to \mathbb{R}$ be an inner product, and suppose that with respect to this product the column vectors

$$a_i := \begin{pmatrix} a_{1i} \\ \vdots \\ a_{ni} \end{pmatrix} \in \mathbb{R}^n$$

are perpendicular to b. Suppose further that $b \neq 0$. Show that the system is not solvable.

7.2P: Exercise 7.2 (for mathematicians)

7.3P: Provide the *mathematical* reasons why it is impossible to determine the masses by the method described in Section 7.6, if none of these (not even M_0) is known beforehand.

Chapter 8
Euclidean Vector Spaces

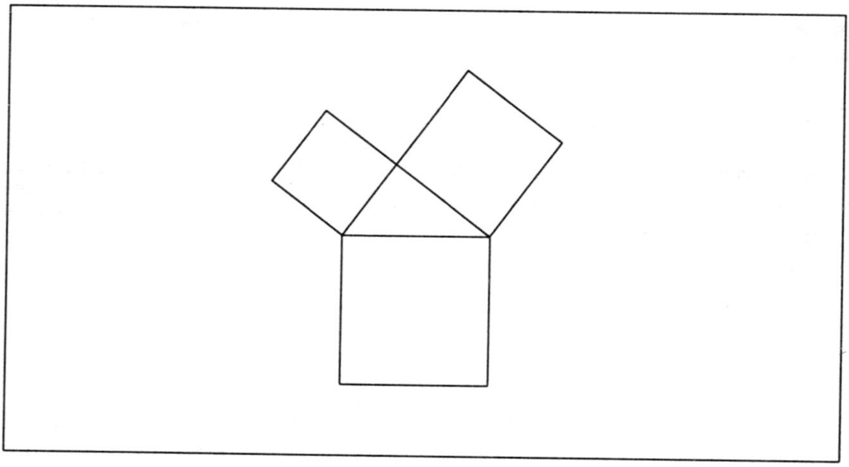

8.1 Inner Products

If one wants to study geometric problems, in which lengths and angles play a role, then the data given by a vector space no longer suffice, and one has to equip the vector space with "additional" structure. The additional structure one needs for metric or *Euclidean* geometry in a real vector space is the inner or "scalar" product. By this we do not mean the scalar multiplication $\mathbb{R} \times V \to V$, but a new kind of composition, $V \times V \to \mathbb{R}$.

Definition: Let V be a real vector space. An ***inner product*** in V is a map $V \times V \to \mathbb{R}$, $(x, y) \mapsto \langle x, y \rangle$ with the following properties:

(i) Bilinearity: for each $x \in V$ the maps

$$\begin{aligned} \langle \cdot\,, x \rangle : V &\longrightarrow \mathbb{R} \\ v &\longmapsto \langle v, x \rangle \end{aligned} \quad \text{and} \quad \begin{aligned} \langle x\,, \cdot \rangle : V &\longrightarrow \mathbb{R} \\ v &\longmapsto \langle x, v \rangle \end{aligned}$$

are linear.

(ii) Symmetry: $\langle x, y \rangle = \langle y, x \rangle$ for all $x, y \in V$

(iii) Positive definiteness: $\langle x, x \rangle > 0$ for all $x \neq 0$.

In short one says that $\langle\, ,\, \rangle$ is a positive definite symmetric bilinear form on V. Because of the symmetry, it would have been enough to give just one of the linearity conditions in (i).

Definition: A ***Euclidean vector space*** is a pair $(V, \langle\, ,\, \rangle)$ consisting of a real vector space V and an inner product $\langle\, ,\, \rangle$ on V.

Without further ado we shall speak of a "Euclidean vector space V" — a double meaning for V — similar to the way in which we always write V instead of $(V, +, \cdot)$.

Important example: The inner product defined by

$$\begin{aligned} \langle\, ,\, \rangle : \mathbb{R}^n \times \mathbb{R}^n &\longrightarrow \mathbb{R} \\ (x, y) &\longmapsto x_1 y_1 + \cdots + x_n y_n \end{aligned}$$

is called the usual or ***standard inner product*** on $\mathbb{R}^n$.

Here is another, quite different, example: let V be the real vector space of continuous functions from $[-1, 1]$ to $\mathbb{R}$. Then, for example,

$$\langle f, g \rangle := \int_{-1}^{1} f(x) g(x) dx$$

defines an inner product. In fact, for *any* real vector space one can find an inner product, and not only one, but infinitely many (unless V={0}). In the case of $\mathbb{R}^n$, for instance, the standard inner product is only the most obvious. For example, $(x, y) \mapsto \langle Ax, Ay \rangle$ fulfills the three conditions for an inner product for any fixed invertible $n \times n$ matrix A.

Definition: If $(V, \langle\, ,\, \rangle)$ is a Euclidean vector space and $x \in V$, then by the ***norm*** of x one understands the real number $\|x\| := \sqrt{\langle x, x \rangle} \geq 0$.

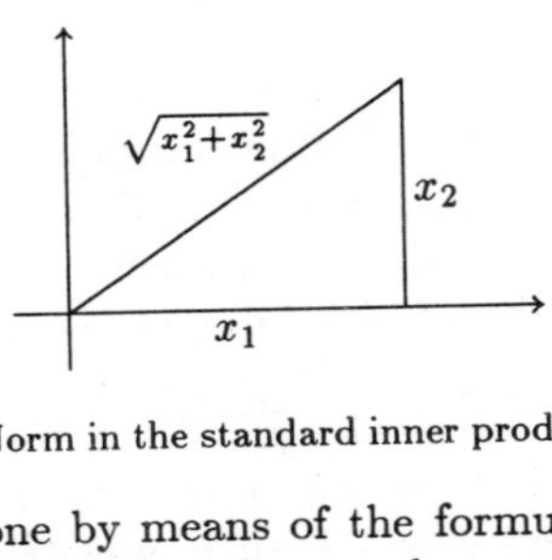

Fig. 47. Norm in the standard inner product on $\mathbb{R}^2$

In $\mathbb{R}^n$ with the usual inner product we have $\|x\| = \sqrt{x_1^2 + \cdots + x_n^2}$. Next we would like to define the *angle* between two nonzero elements x, y in a Euclidean vector space V. (Note: not "determine" or "calculate," but first of all define!) This will be done by means of the formula $\langle x, y \rangle = \|x\|\,\|y\| \cos \alpha(x, y)$, $0 \leqq \alpha(x, y) \leqq \pi$. But before this can be used as a definition for $\alpha(x, y)$, we must show that $-1 \leq \frac{\langle x, y \rangle}{\|x\|\|y\|} \leq 1$. This is called the

Cauchy-Schwarz inequality.

Theorem 1 (Cauchy-Schwarz inequality): In each Euclidean vector space V we have $|\langle x,y\rangle| \leqq \|x\|\,\|y\|$ for all $x,y \in V$.

PROOF: For $y=0$ this is trivial. Let $y \neq 0$. We now apply a small trick: set $\lambda := \frac{\langle x,y\rangle}{\|y\|^2} \in \mathbb{R}$ and calculate

$$\begin{aligned} 0 \leqq \langle x-\lambda y, x-\lambda y\rangle &= \langle x,x\rangle - 2\lambda\langle x,y\rangle + \lambda^2\langle y,y\rangle \\ &= \|x\|^2 - 2\frac{\langle x,y\rangle^2}{\|y\|^2} + \frac{\langle x,y\rangle^2}{\|y\|^2} \\ &= \|x\|^2 - \frac{\langle x,y\rangle^2}{\|y\|^2}. \end{aligned}$$

Therefore, $\langle x,y\rangle^2 \leqq \|x\|^2\|y\|^2$, and as required $|\langle x,y\rangle| \leqq \|x\|\,\|y\|$. □

Theorem 2: If V is a Euclidean vector space, the norm $\|..\| : V \to \mathbb{R}$ has the following properties:

$$\begin{array}{ll} \text{(i)} & \|x\| \geqq 0 \quad \text{for all } x. \\ \text{(ii)} & \|x\| = 0 \iff x = 0. \\ \text{(iii)} & \|\lambda x\| = |\lambda| \cdot \|x\| \quad \text{for all } x \in V,\ \lambda \in \mathbb{R}. \\ \text{(iv)} & \|x+y\| \leqq \|x\| + \|y\| \quad \text{for all } x,y \in V. \end{array}$$

Property (iv) of the norm is called the ***triangle inequality.***

PROOF: (i)–(iii) are clear from the definition. For the triangle inequality we have

$$\begin{aligned} (\|x\|+\|y\|)^2 &= \|x\|^2 + 2\|x\|\|y\| + \|y\|^2 \\ &\geqq \|x\|^2 + 2\langle x,y\rangle + \|y\|^2 = \|x+y\|^2 \end{aligned}$$

by the Cauchy-Schwarz inequality. Therefore, $\|x\|+\|y\| \geqq \|x+y\|$. □

Inequality (iv) is called the "triangle inequality" for the following reason. If $a,b,c \in V$ one takes $\|a-b\|, \|a-c\|$, and $\|b-c\|$ to be the lengths of the sides of the triangle with vertices a,b,c. The triangle inequality applied to $x = a-b$, $y = b-c$ then says that $\|a-c\| \leq \|a-b\| + \|b-c\|$, i.e., the length of one side is less than or equal to the sum of the lengths of the other two sides (see Fig. 48). But note that the statement that the length of the segment from a to b equals $\|a-b\|$ is not a theorem but a *definition*.

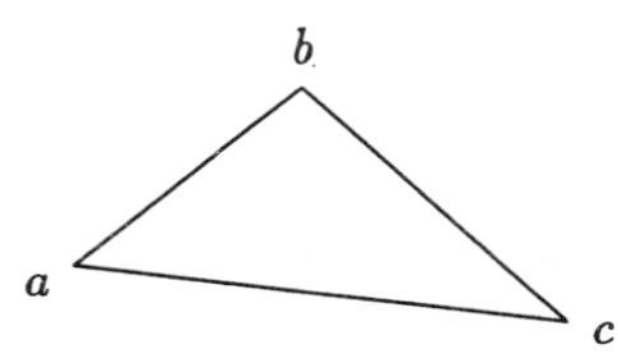

Fig. 48. In a Euclidean vector space the triangle inequality $\|a-c\| \leq \|a-b\| + \|b-c\|$ holds true.

Definition: For two nonzero elements x, y in the Euclidean vector space V, one defines the angle $\alpha(x, y)$ formed by x and y by

$$\cos\alpha(x, y) = \frac{\langle x, y\rangle}{\|x\|\,\|y\|}, \quad 0 \leqq \alpha(x, y) \leqq \pi.$$

(Because of the Cauchy-Schwarz inequality, we have

$$-1 \leqq \frac{\langle x, y\rangle}{\|x\|\,\|y\|} \leqq 1,$$

and since the cosine defines a *bijective* map $\cos : [0, \pi] \to [-1, 1]$, the angle $\alpha(x, y)$ is welldefined by the above requirements.)

8.2 Orthogonal Vectors

Definition: Two elements v, w of a Euclidean vector space are said to be ***orthogonal*** or perpendicular to each other (written $v \perp w$), if $\langle v, w\rangle = 0$.

The definition of $\alpha(v, w)$ for nonzero vectors implies that, if $\langle v, w\rangle = 0$, then the angle between v and w is 90^0.

The label "orthogonal" is applicable to many concepts associated with Euclidean vector spaces; among others we have "orthogonal complement," "orthogonal projection," "orthogonal transformation," and "orthogonal matrix," see below. Put very generally, orthogonality means "compatibility" with the inner product. But of course this vague general explanation is not meant to replace the precise individual definitions.

Definition: If M is a subset of the Euclidean vector space V, then $M^\perp := \{v \in V \mid v \perp u \text{ for all } u \in M\}$ is called the *orthogonal complement* of M. Instead of "$v \perp u$ for all $u \in M$" one can use the shorter notation $v \perp M$.

Remark: $M^\perp$ is a subspace of V.

PROOF: $M^\perp \neq \varnothing$, since $0 \perp M$, and if $x, y \in M^\perp$ and $\lambda \in \mathbb{R}$, then $x + y \in M^\perp$ and $\lambda x \in M^\perp$ follows immediately from the linearity of $\langle \cdot\,, u\rangle : V \to \mathbb{R}$. □

Definition: An r-tuple $(v_1, \ldots, v_r)$ of vectors in a Euclidean vector space is said to be *orthonormal* or an *orthonormal system* if $\|v_i\| = 1$, $i = 1, \ldots, r$, and $v_i \perp v_j$ for $i \neq j$. (Expressed differently: $\langle v_i, v_j\rangle = \delta_{ij}$).

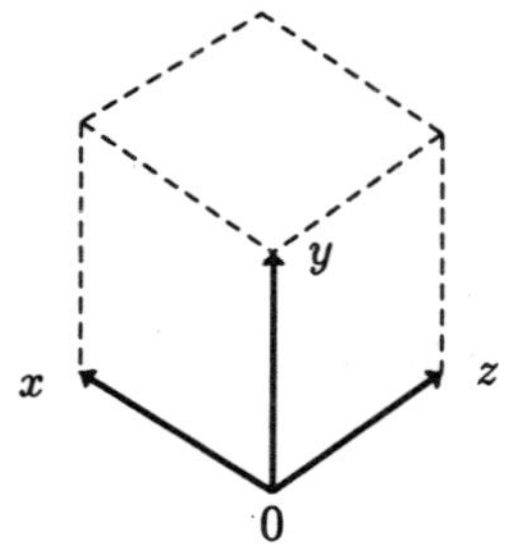

Fig. 49. An orthonormal system (x,y,z)

If, for instance, x, y, z are the three "edge vectors" of a cube sitting at the origin of $\mathbb{R}^3$ with edge length equal to 1, then (x, y, z) is an orthonormal system (pairwise mutually perpendicular vectors of length 1). See the diagram in Fig. 49. (But the pair (x, y) and even (z) alone are also orthonormal systems: an orthonormal system need not be a basis.) In arguments with an orthonormal system $v_1, \ldots, v_r$, one repeatedly uses the same little trick, namely to apply the operation $\langle \cdot\, , v_i \rangle$ to both sides of an equation. Look at its use in the proofs of the following three lemmas.

Lemma 1: An orthonormal system is always linearly independent.

PROOF: Let $(v_1, \ldots, v_r)$ be orthonormal and $\lambda_1 v_1 + \cdots + \lambda_r v_r = 0$. Then $\langle \lambda_1 v_1 + \cdots + \lambda_r v_r, v_i \rangle = \lambda_i \langle v_i, v_i \rangle = \lambda_i = 0$ for $i = 1, \ldots, r$. □

Lemma 2: (Expansion with respect to an orthonormal basis): If $(v_1, \ldots, v_n)$ is an orthonormal *basis* of V, then for each $v \in V$ we have the "expansion formula"

$$v = \sum_{i=1}^{n} \langle v, v_i \rangle v_i.$$

PROOF: We know that v is expressible as $v = c_1 v_1 + \cdots + c_n v_n$, because $(v_1, \ldots, v_n)$ is a basis. If we apply $\langle \cdot\, , v_i \rangle$ to both sides of this equation, we obtain $\langle v, v_i \rangle = c_i \langle v_i, v_i \rangle = c_i$ for each $i = 1, \ldots, n$. □

Lemma 3: If $(v_1, \ldots, v_r)$ is an orthonormal system in V and $U := L(v_1, \ldots, v_r)$ denotes the subspace spanned by the orthonormal system, then each $v \in V$ can be uniquely expressed as a sum $v = u + w$ with $u \in U$ and $w \in U^\perp$. Indeed

$$u = \sum_{i=1}^{r} \langle v, v_i \rangle v_i$$

and hence $w = v - \sum_{i=1}^{r} \langle v, v_i \rangle v_i$.

PROOF: The fact that v can be expressed in *at most* one way as a sum $v = u + w$ with $u \in U$ and $w \in U^\perp$ follows from the positive definiteness of the inner product, because if $v = u + w = u' + w'$ with $u, u' \in U$ and $w, w' \in U^\perp$ then $(u - u') + (w - w') = 0$ and $\langle u - u', w - w' \rangle = 0$, hence $\langle u - u', u - u' \rangle = 0$. This implies that $u - u' = 0$, and as a consequence $w - w' = 0$ also.

This holds for any subspace U of V, even if it were not generated by a finite orthonormal system. But we use our assumption in the existence part of the proof.

Any $u \in U$ can now be written as $u = c_1 v_1 + \cdots + c_r v_r$, and all we have to do is to find coefficients $c_1, \ldots, c_r$ such that the vector $w := v - u$ is in $U^\perp$. But $w \in U^\perp$ means $\langle w, v_i \rangle = 0$ for $i = 1, \ldots, r,$, that is, if $\langle v, v_i \rangle - \langle u, v_i \rangle = 0$, or $c_i = \langle v, v_i \rangle$. See Fig. 50. □

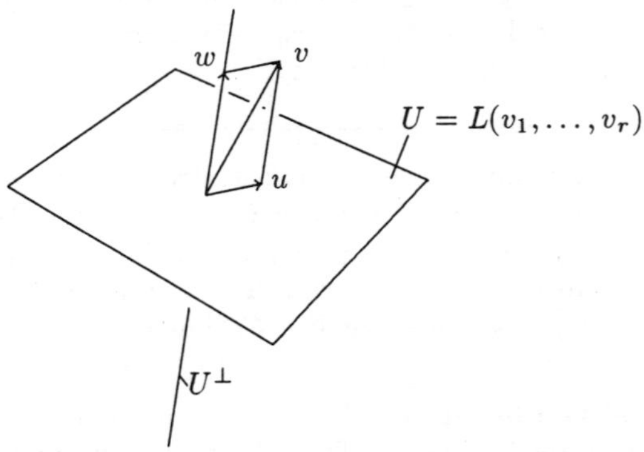

Fig. 50. Splitting a vector into its U and $U^\perp$ components

In these three lemmas we have always assumed that the orthonormal system is given. But how does one obtain orthonormal systems? For this there exists, for example, the *Gram-Schmidt orthonormalization process*, in which one successively changes an arbitrary linearly independent r-tuple of vectors $v_1, \ldots, v_r$ into an orthonormal system $\widetilde{v}_1, \ldots, \widetilde{v}_r$, and in such a way that the first k vectors of each system always span the same space:

$$U_k := L(v_1, \ldots, v_k) = L(\widetilde{v}_1, \ldots, \widetilde{v}_k)$$

for $k = 1, \ldots, r$. Of course one begins by "normalizing" the first vector v_1, thus $\widetilde{v}_1 := v_1 / \|v_1\|$. But it does not suffice to normalize all of the v_i, since then one would have achieved length 1 but not mutual orthogonality. Rather, before normalization we must replace the vector v_{k+1}, as in Lemma 3, by its component $w_{k+1} \in U_k^\perp$ perpendicular to U_k. By the inductive assumption, U_k is spanned by the orthonormal system $(\widetilde{v}_1, \ldots, \widetilde{v}_r)$, and Lemma 3 gives us the explicit calculation formula

$$w_{k+1} := v_{k+1} - \sum_{i=1}^{k} \langle v_{k+1}, \widetilde{v}_i \rangle \widetilde{v}_i$$

for a vector that is perpendicular to U_k, and which together with $\widetilde{v}_1, \ldots, \widetilde{v}_k$ spans the space U_{k+1}. We need only to normalize this vector in order to obtain $\widetilde{v}_{k+1}$.

Gram-Schmidt Orthonormalization Process: If $(v_1, \ldots, v_r)$ is a linearly independent r-tuple of vectors in a Euclidean vector space V, then by defining $\widetilde{v}_1 := v_1/\|v_1\|$ and using the recursion formula

$$\widetilde{v}_{k+1} := \frac{v_{k+1} - \sum_{i=1}^{k} \langle v_{k+1}, \widetilde{v}_i \rangle \widetilde{v}_i}{\|v_{k+1} - \sum_{i=1}^{k} \langle v_{k+1}, \widetilde{v}_i \rangle \widetilde{v}_i\|}$$

for $k = 1, \ldots, r-1$, we obtain an orthonormal system $(\widetilde{v}_1, \ldots, \widetilde{v}_r)$. The first k vectors $\widetilde{v}_1, \ldots, \widetilde{v}_k$ span the same subspace as the original $v_1, \ldots, v_k$. In particular, if $(v_1, \ldots, v_n)$ has been a basis of V, then $(\widetilde{v}_1, \ldots, \widetilde{v}_n)$ is an orthonormal *basis* of V.

In particular, in any finite-dimensional Euclidean vector space it is possible to find *some* orthonormal basis, and therefore the unique decomposition of $v \in V$ into U and $U^\perp$ components (Lemma 3) is indeed applicable for *each* finite dimensional subspace U of a Euclidean vector space. One calls the map $v \mapsto u$ *orthogonal projection on* U, as Fig. 51 illustrates.

Corollary and Definition: Let V be a Euclidean vector space and U be a finite-dimensional subspace. Then there exists a unique linear map $P_U : V \longrightarrow U$ with $P_U|U = \mathrm{Id}_U$ and $\mathrm{Ker}(P_U) = U^\perp$. This map P_U is called the ***orthogonal projection*** onto U.

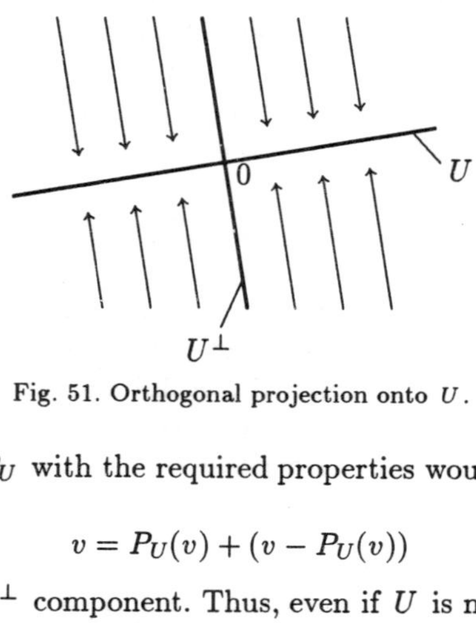

Fig. 51. Orthogonal projection onto U.

PROOF: Each map P_U with the required properties would induce a decomposition

$$v = P_U(v) + (v - P_U(v))$$

of v into a U and $U^\perp$ component. Thus, even if U is not finite-dimensional, there exists *at most* one such map. If we now choose an orthonormal basis $(v_1, \ldots, v_r)$ in U, we are prompted by the construction in Lemma 3 to define

$$P_U(v) := \sum_{i=1}^{r} \langle v, v_i \rangle v_i,$$

and this linear map $P_U : V \to U$ clearly has the right properties. □

Corollary: If U is a subspace of a finite-dimensional Euclidean vector space V, and $U^\perp = 0$, then $U = V$.

PROOF: $P_U : V \to U$ is always surjective, and since $\text{Ker}(P_U) = U^\perp = 0$, also injective. Hence P_U is bijective, and therefore $U = V$ follows from $P_U|U = \text{Id}_U$. □

In the next section we will introduce the notion of *orthogonal maps*. From a consistent terminology you might expect orthogonal projections to be orthogonal. But alas! Instead of defending the terminology I shall take cover, declaring that I am not to blame.

8.3 Orthogonal Maps

Definition: Let V and V' be Euclidean vector spaces. A linear map $f : V \to V'$ is said to be ***orthogonal*** or ***isometric*** if

$$\langle f(v), f(w)\rangle = \langle v, w\rangle$$

for all $v, w \in V$.

Fact: An orthogonal map is always injective, since if $v \in \text{Ker} f$, we have $\langle 0, 0\rangle = \langle v, v\rangle$. Thus for *finite-dimensional* Euclidean spaces, orthogonal endomorphisms are necessarily automorphisms.

This follows from the dimension formula $\dim \text{Ker} f + \dim \text{Im} f = \dim V$, showing that f is surjective, too, and hence an isomorphism.

Definition: We denote the set of orthogonal isomorphisms of a Euclidean vector space V by $O(V)$. Instead of $O(\mathbb{R}^n)$, where $\mathbb{R}^n$ is equipped with the standard inner product, we use the shorter form $O(n)$. If in the usual way the elements of $O(n)$ are considered as $n \times n$ real matrices, $O(n) \subset M(n \times n, \mathbb{R})$, these matrices are called ***orthogonal matrices***.

Remark: Let V, V' be Euclidean vector spaces, and let $(v_1, \dots, v_n)$ be an orthonormal basis of V. Then a linear map $f : V \to V'$ is orthogonal if and only if $(f(v_1), \dots, f(v_n))$ is an orthonormal system in V'.

PROOF: If f is orthogonal, then $\langle f(v_i), f(v_j)\rangle = \langle v_i, v_j\rangle = \delta_{ij}$. Conversely, if we *assume* that $\langle f(v_i), f(v_j)\rangle = \delta_{ij}$, then it follows for $v := \Sigma\lambda_i v_i$ and $w := \Sigma\mu_j v_j \in V$, that $\langle f(v), f(w)\rangle = \langle f(\Sigma\lambda_i v_i),\ f(\Sigma\mu_j v_j)\rangle = \langle \Sigma\lambda_i f(v_i), \Sigma\mu_j f(v_j)\rangle = \Sigma_i\Sigma_j\lambda_i\mu_j\delta_{ij} = \Sigma_i\Sigma_j\lambda_i\mu_j\langle v_i, v_j\rangle = \langle\Sigma\lambda_i v_i, \Sigma\mu_j v_j\rangle = \langle v, w\rangle$. □

Corollary: A matrix $A \in M(n \times n, \mathbb{R})$ is orthogonal if and only if the columns (images of the unit vectors!) form an orthonormal system with respect to the usual inner product in $\mathbb{R}^n$, that is, if $A^tA = E$ holds.

If we denote the columns of A by $s_1, \ldots, s_n$, the s_i are the rows of A^t, and the element in the (i, j) position of the product A^tA thus equals $\langle s_i, s_j \rangle$:

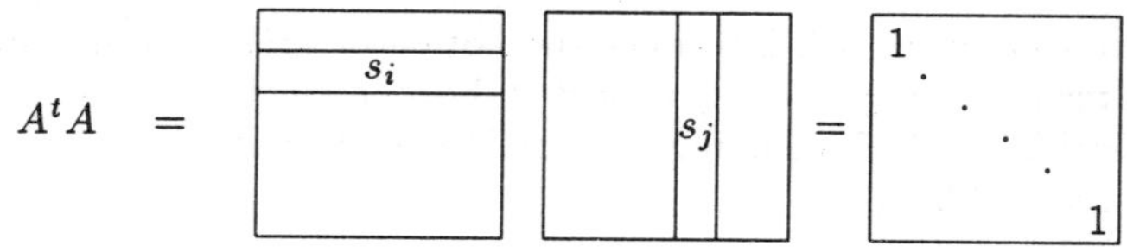

Using our knowledge of invertible matrices, we can draw some immediate consequences of this.

Fact: For $A \in M(n \times n, \mathbb{R})$, the following conditions are equivalent:
(i) A is orthogonal.
(ii) The columns form an orthonormal system.
(iii) $A^tA = E$.
(iv) A is invertible and $A^{-1} = A^t$.
(v) $AA^t = E$.
(vi) The rows form an orthonormal system.

Fact and Definition: It follows from $A^tA = E$ that $\det A^t \cdot \det A = (\det A)^2 = 1$, and hence that $\det A = \pm 1$ for all $A \in O(n)$. A matrix $A \in O(n)$ is called a ***special orthogonal matrix*** if $\det A = +1$. The set of special orthogonal matrices is denoted by $SO(n)$.

8.4 Groups

If we compose two orthogonal matrices $A, B \in O(n)$ by means of matrix multiplication, we again obtain an orthogonal matrix $AB \in O(n)$. And $(AB)C = A(BC)$ for all $A, B, C \in O(n)$, as indeed for all matrices in $M(n \times n, \mathbb{R})$. Furthermore, $O(n)$ contains a multiplicatively "neutral" element, namely the unit matrix E, and for each $A \in O(n)$ there exists an orthogonal matrix A^{-1} in $O(n)$ with $AA^{-1} = A^{-1}A = E$, the inverse matrix to A. Because of these properties the pair

$$(O(n), O(n) \times O(n) \to O(n)),$$

consisting of the set $O(n)$ and the operation of matrix multiplication, is called a ***group***. The concept of a group is of fundamental importance in all mathematics, not only in algebra, even though it is an "algebraic" concept. You will see how, in the material already covered, we have met a whole collection of groups.

Definition: A ***group*** is a pair $(G,\cdot)$ consisting of a set G and a map

$$\begin{aligned} \cdot : G \times G &\longrightarrow G \\ (a,b) &\longmapsto ab \end{aligned}$$

such that the following three axioms are satisfied:

(1) Associativity: $(ab)c = a(bc)$ for all $a,b,c \in G$,

(2) Existence of a neutral element: there exists an $e \in G$ with $ae = ea = a$ for all $a \in G$.

(3) Existence of inverses: for each $a \in G$ there exists an $a^{-1} \in G$ with $aa^{-1} = a^{-1}a = e$.

Note that an initial axiom (0): "for $a,b \in G$ we also have $ab \in G$" only becomes superfluous, because it is already included in the assumption that "$\cdot$" is a map from $G \times G \to G$. If, however, as is often the case, one starts out from an operation into a larger set only *containing* G as a subset, then one actually has to check (0). As an example, consider the set $G = \{x \in \mathbb{R} \mid \frac{1}{2} < x < 2\}$ with the operation of multiplication of real numbers. Even though axioms (1), (2), and (3) hold, $(G,\cdot)$ fails to be a group, because the operation does not remain "inside G": $\frac{3}{2} \cdot \frac{3}{2} = \frac{9}{4} > 2$.

Definition: If a group $(G,\cdot)$ has the additional property of being "commutative," that is

(4) $ab = ba$ for all $a,b \in G$,

then $(G,\cdot)$ is called an ***abelian group***.

Remark: A group $(G,\cdot)$ contains a unique neutral element e, and the inverse element a^{-1} is unique for each $a \in G$.

PROOF: If $ea = ae = e'a = ae' = a$ for all $a \in G$, then in particular $ee' = e$, because e' is neutral, and $ee' = e'$, because e is neutral. Hence $e = e'$. If $a \in G$ is given and $ba = ab = ca = ac = e$, then $c = c(ab) = (ca)b = eb = b$. □

Notation: As you will expect, one just writes G instead of $(G,\cdot)$. Frequently the neutral element is denoted by 1. For abelian groups one often writes the operation as "addition," that is, as

$$(a,b) \longmapsto a+b$$

instead of $(a,b) \mapsto ab$, and correspondingly then denotes the neutral element by 0 and the element inverse to a by $-a$. In principle, one could do this for any group, but for nonabelian groups it is not customary.

Examples of groups:

1. $(\mathbb{Z}, +)$ is an abelian group.
2. $(\mathbb{R}, +)$ is an abelian group.
3. $(\mathbb{R}\setminus\{0\}, \cdot)$ is an abelian group.
4. If $(\mathbb{F}, +, \cdot)$ is a field, then $(\mathbb{F}, +)$ is an abelian group.
5. If $(\mathbb{F}, +, \cdot)$ is a field, then $(\mathbb{F}\setminus\{0\}, \cdot)$ is an abelian group.
6. If $(V, +, \cdot)$ is a vector space over $\mathbb{F}$, then $(V, +)$ is an abelian group.
7. If M is a set, $\mathrm{Bij}(M)$ the set of bijective maps $f: M \to M$ and $\circ$ is the symbol for composition of maps, then $(\mathrm{Bij}(M), \circ)$ is a group. The neutral element is the identity.

Frequently you will come across groups in which the elements are special bijective self-maps of a set M, and the operation is the composition of these maps, that is, "subgroups" of $\mathrm{Bij}(M)$, so to say. See the following examples 8–12.

Fact: If $G \subset Bij(M)$ and $\mathrm{Id}_M \in G$ and if $f \circ g \in G$, $f^{-1} \in G$ for all $f, g \in G$, then $(G, \circ)$ is a group.

Further examples: In the following examples of groups, the operation is defined by the composition of maps:

8. $GL(V)$, the group of automorphisms of a vector space V over $\mathbb{F}$.
9. $GL(n, \mathbb{F}) := GL(\mathbb{F}^n)$, the group of invertible $n \times n$ matrices over $\mathbb{F}$, called the *general linear group*.
10. $SL(n, \mathbb{F}) = \{A \in GL(n, \mathbb{F}) \mid \det A = 1\}$, the *special linear group*.
11. $O(n)$, the *orthogonal group*.
12. $SO(n)$, the *special orthogonal group*.

8.5 Test

(1) An inner product on a real vector space is a map

- ☐ $\langle\, , \rangle : V \times V \to \mathbb{R}$
- ☐ $\langle\, , \rangle : V \times V \to V$
- ☐ $\langle\, , \rangle : \mathbb{R} \times V \to V$

(2) Positive definiteness of the inner product means that

- ☐ $\langle x, y\rangle > 0 \Rightarrow x = y$.
- ☐ $\langle x, x\rangle > 0 \Rightarrow x \neq 0$.
- ☐ $\langle x, x\rangle > 0$ for all $x \in V$, $x \neq 0$.

(3) Which of the following statements is (or are) correct?

- ☐ If $\langle\ ,\ \rangle : \mathbb{R}^n \times \mathbb{R}^n \to \mathbb{R}$ is an inner product on the real vector space $\mathbb{R}^n$, then $\langle x, y\rangle = x_1y_1 + \cdots + x_ny_n$ for all $x, y \in \mathbb{R}^n$.
- ☐ If $\langle\ ,\ \rangle : \mathbb{R}^n \times \mathbb{R}^n \to \mathbb{R}^n$ is an inner product on the real vector space $\mathbb{R}^n$, then $\langle x, y\rangle = (x_1y_1, \ldots, x_ny_n)$ for all $x, y \in \mathbb{R}^n$.
- ☐ If one defines $\langle x, y\rangle := x_1y_1 + \cdots + x_ny_n$ for all $x, y \in \mathbb{R}^n$, then one obtains an inner product on $\mathbb{R}^n$.

(4) By the orthogonal complement $U^\perp$ of a subspace U of the Euclidean vector space V, one understands

- ☐ $U^\perp := \{u \in U \mid u \perp U\}$.
- ☐ $U^\perp := \{x \in V \mid x \perp U\}$.
- ☐ $U^\perp := \{x \in V \mid x \perp U \text{ and } \|x\| = 1\}$.

(5) Let $V = \mathbb{R}^2$ with the standard inner product. Which of the following tuples of elements of V forms an orthonormal basis?

- ☐ $((1,-1),(-1,-1))$
- ☐ $((-1,0),(0,-1))$
- ☐ $((1,0),(0,1),(1,1))$

(6) Which of the following conditions on a linear map $f : V \to W$ of one Euclidean space into another is equivalent to f being orthogonal?

- ☐ $\langle f(x), f(y)\rangle > 0$ for all $x, y \in V$.
- ☐ $\langle x, y\rangle = 0 \iff \langle f(x), f(y)\rangle = 0$.
- ☐ $\|f(x)\| = \|x\|$ for all $x \in V$.

(7) For which subspaces $U \subset V$ is the orthogonal projection $P_U : V \to U$ an orthogonal map?

☐ for each U ☐ only for $U = V$ ☐ only for $V = \{0\}$

(8) Which of the following matrices is (or are) orthogonal?

☐ $\begin{pmatrix} 0 & 1 \\ -1 & 0 \end{pmatrix}$ ☐ $\begin{pmatrix} 1 & -1 \\ -1 & 1 \end{pmatrix}$ ☐ $\begin{pmatrix} 1 & 1 \\ 0 & 1 \end{pmatrix}$

(9) Which of the following arguments correctly explains why $(\mathbb{N}, +)$ fails to be a group?

- ☐ For natural numbers we have $n + m = m + n$, but this is not one of the group axioms, so $(\mathbb{N}, +)$ fails to be a group.
- ☐ The operation $\mathbb{N} \times \mathbb{N} \to \mathbb{N}$, $(n, m) \mapsto n + m$, is not defined for all integers, because the negative numbers do not belong to $\mathbb{N}$. Therefore, $(\mathbb{N}, +)$ fails to be a group.
- ☐ The third group axiom (existence of inverses) is not satisfied, since, for example, for $1 \in \mathbb{N}$ there exists no $n \in \mathbb{N}$ with $1 + n = 0$. Therefore, $(\mathbb{N}, +)$ fails to be group.

(10) For $k > 0$ we have

☐ $SO(2k) \subset O(k)$.

☐ $SO(2k) \subset O(2k)$, but $SO(2k) \neq O(2k)$.

☐ $SO(2k) = O(2k)$, because $(-1)^{2k} = 1$.

8.6 Remark on the Literature

If circumstances allow, namely if the instructor is not limited in his decisions either by powerful sevice requirements or by the syllabus, he will arrange his course to reach as quickly as possible the goal he has set for himself and his students. In particular, this can give a sharp twist to the course in the direction of algebra. You will first realize that you are in such a course if the fundamental algebraic structures such as groups, rings, and fields are formally introduced at the outset, and then (an unmistakable sign) instead of vector spaces over fields, the more general *modules* over *rings* appear.

In this situation you can only use the present text as a supplement or as contrasting material, and although it grieves me to lose you as a customer or even reader, it is my duty to advise you to go and find yourself a more suitable, more algebraically conceived textbook.

8.7 Exercises

Exercises for mathematicians

8.1: Prove Pythagoras's theorem: if the three points a, b, c in a Euclidean vector space form a right-angled triangle, that is if $a - c \perp b - c$, then $\|a - c\|^2 + \|b - c\|^2 = \|a - b\|^2$ (see Fig. 52).

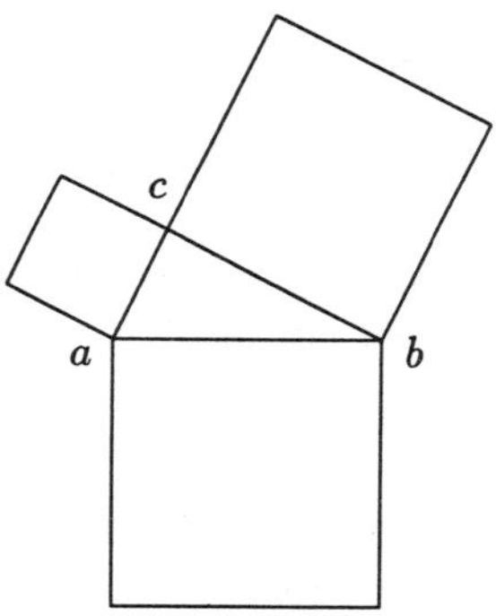

Fig. 52. Is $\|a-c\|^2+\|b-c\|^2=\|a-b\|^2$ an honest version of the Pythagorean theorem?

8.2: Give $\mathbb{R}^3$ the inner product $\langle x, y\rangle := \sum_{i,j=1}^{3} a_{ij}x_iy_j$, where

$$A = \begin{pmatrix} 2 & 1 & 0 \\ 1 & 2 & 0 \\ 0 & 1 & 4 \end{pmatrix}.$$

(That this is actually an inner product is not part of the exercise, and may be assumed.) Calculate the cosines of the angles between the canonical unit vectors in $\mathbb{R}^3$.

8.3: Show that the 2×2 matrices $A \in O(2)$, whose coefficients take only the values $0, +1$ and -1, form a nonabelian group.

Three $*$-exercises

8.1*: For $x := (x_1, \ldots, x_n) \in \mathbb{R}^n, n \geq 2$, define $|x| := \max_i |x_i|$. Show that there exists no inner product $\langle\ ,\ \rangle$ on $\mathbb{R}^n$, for which $\langle x, x\rangle = |x|^2$ for all $x \in \mathbb{R}^n$.

8.2*: Let V be the real vector space of all bounded real sequences,

$$V := \{(x_i)_{i=1,2\ldots} \mid x_i \in \mathbb{R} \text{ and } \exists c \in \mathbb{R} \text{ with } |x_i| \leq c \text{ for all } i\}.$$

Then

$$\langle x, y\rangle := \sum_{n=1}^{\infty} \frac{x_n y_n}{n^2}$$

obviously defines an inner product on V. Find a proper vector subspace $U \subsetneqq V$ with $U^{\perp} = \{0\}$.

8.3*: Let M be a set. Show that if $(\mathrm{Bij}(M), \circ)$ is abelian, then M has fewer than three elements.

Exercises for physicists

8.1P: Exercise 8.1 (for mathematicians)

8.2P: Use the Gram-Schmidt orthogonalization process to find an orthonormal basis for the subspace

$$U := L((-3,-3,3,3), (-5,-5,7,7), (4,-2,0,6))$$

of $\mathbb{R}^4$, where $\mathbb{R}^4$ is given the usual inner product.

8.3P: Let V be a Euclidean vector space. Show

(a) If $\varphi : \mathbb{R} \to V$ and $\psi : \mathbb{R} \to V$ are differentiable maps, then $\langle\varphi, \psi\rangle : \mathbb{R} \to \mathbb{R}$, $t \mapsto \langle\varphi(t), \psi(t)\rangle$ is also differentiable, and we have

$$\langle\varphi, \psi\rangle^{\boldsymbol{\cdot}}(t) = \langle\dot{\varphi}(t), \psi(t)\rangle + \langle\varphi(t), \dot{\psi}(t)\rangle.$$

(b) If $\varphi : \mathbb{R} \to V$ is differentiable and $\|\varphi\|$ is constant, then $\varphi(t) \perp \dot{\varphi}(t)$ for all $t \in \mathbb{R}$.

Hint: Differentiability of a map ("curve") of $\mathbb{R}$ into a Euclidean vector space is defined exactly as for real-valued functions: the map $\varphi : \mathbb{R} \to V$ is said to be differentiable at t_0, if the limit

$$\lim_{t \to t_0} \frac{\varphi(t) - \varphi(t_0)}{t - t_0} =: \dot{\varphi}(t_0)$$

exists. This means that there exists a vector $v_0 \in V$ (which will be $\dot{\varphi}(t_0)$), such that for each $\varepsilon > 0$ there exists $\delta > 0$, such that for all t with $0 < |t - t_0| < \delta$, we have

$$\left\| \frac{\varphi(t) - \varphi(t_0)}{t - t_0} - v_0 \right\| < \varepsilon.$$

CHAPTER 9

Eigenvalues

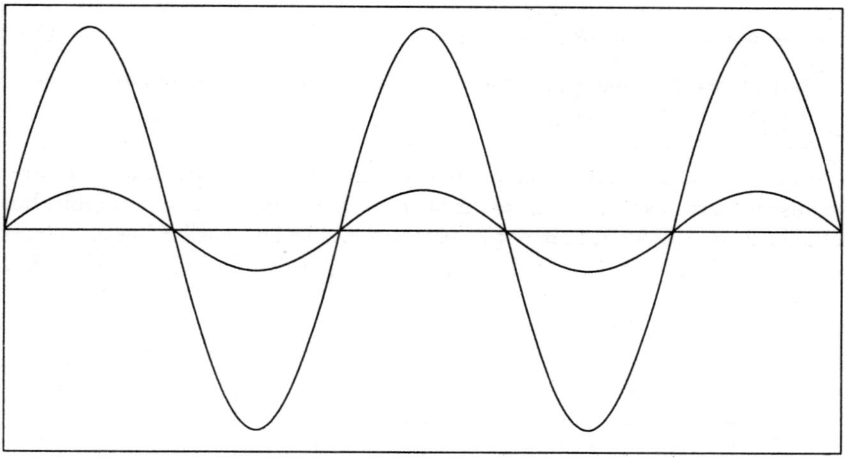

9.1 Eigenvalues and Eigenvectors

Definition: Let V be a vector space over $\mathbb{F}$ and $f : V \to V$ be an endomorphism. By an ***eigenvector*** of f associated with the ***eigenvalue*** $\lambda \in \mathbb{F}$, we mean a vector $v \neq 0$ from V with the property $f(v) = \lambda v$.

Lemma: If A is the matrix of $f : V \to V$ with respect to the basis $(v_1, \ldots, v_n)$ of V, then A is in "diagonal form," that is

$$A = \begin{pmatrix} \lambda_1 & & \\ & \ddots & \\ & & \lambda_n \end{pmatrix}$$

(with zeros away from the diagonal), if and only if v_i is an eigenvector for the eigenvalue λ_i for $i = 1, \ldots, n$.

PROOF: We know that the relation between the endomorphism f and the matrix A is given by the isomorphism $\Phi : \mathbb{F}^n \to V$ mapping e_i onto v_i in the commutative diagram

$$\begin{array}{ccc} V & \xrightarrow{\ f\ } & V \\ \Phi \uparrow \cong & & \Phi \uparrow \cong \\ \mathbb{F}^n & \xrightarrow{\ A\ } & \mathbb{F}^n \end{array}$$

The columns of the diagonal matrix are $\lambda_i e_i$, hence A is in diagonal form if and only if $Ae_i = \lambda_i e_i$. Now use the isomorphism Φ to translate this into an equivalent condition on f:

$$Ae_i = \lambda_i e_i \iff \Phi(Ae_i) = \Phi(\lambda_i e_i) \iff f(\Phi(e_i)) = \lambda_i \Phi(e_i)$$

since $\Phi \circ A = f \circ \Phi$. But $\Phi(e_i) = v_i$! □

Definition: Endomorphisms for which a basis of eigenvectors exists are therefore said to be ***diagonalizable***.

To have a basis of eigenvectors greatly facilitates the study of an operator, because with respect to this basis the operator not only *seems* to be simple, but actually is so. However, it is not always possible to find such a basis. Figures 53, 54, and 55 depict three rather typical examples for $\mathbb{F} := \mathbb{R}$ and $V := \mathbb{R}^2$.

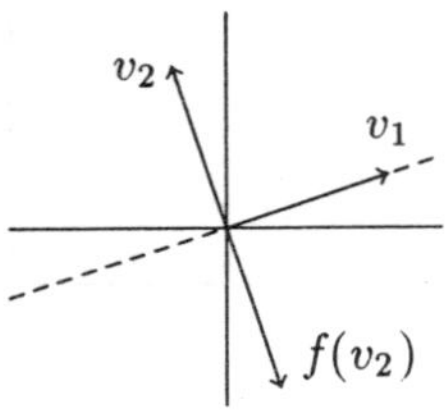

Fig. 53. Reflection

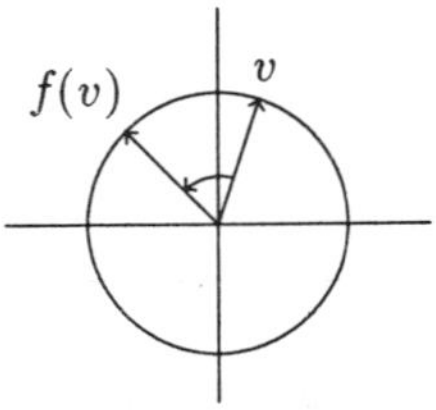

Fig. 54. Rotation through an angle $0<\varphi<\pi$

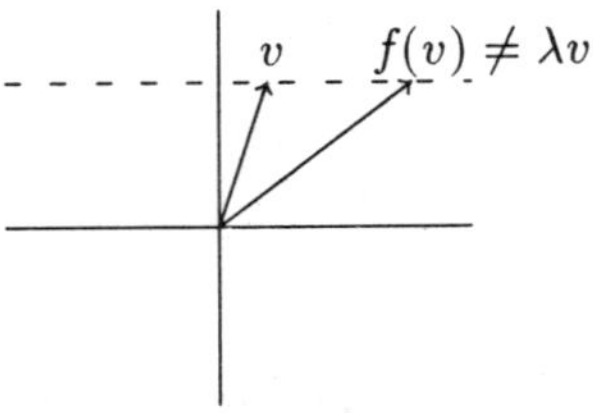

Fig. 55. Shear $\begin{pmatrix} 1 & 1 \\ 0 & 1 \end{pmatrix}$

The reflection does have a basis (v_1, v_2) of eigenvectors, the eigenvalues are $\lambda_1 = 1$ and $\lambda_2 = -1$, and with respect to this basis the associated matrix takes the diagonal form $\begin{pmatrix} 1 & \\ & -1 \end{pmatrix}$. Rotation through an angle $0 < \varphi < \pi$ clearly has no eigenvector, still less a *basis* of eigenvectors. For the shear, only the (nonzero) vectors on the x-axis are eigenvectors (with the eigenvalue $\lambda = 1$). Hence there is no basis of eigenvectors.

Fact: *A vector $v \neq 0$ is an eigenvector of $f : V \to V$ for the eigenvalue $\lambda \in \mathbb{F}$ if and only if $v \in \mathrm{Ker}\,(f - \lambda \mathrm{Id}_V)$.*

An obvious fact, because $(f - \lambda\mathrm{Id})(v) = 0$ just means $f(v) - \lambda v = 0$. But in all its triviality this observation has required a little trick — to *think* of the identity $\mathrm{Id} : V \to V$ and of writing v as $\mathrm{Id}(v)$. A number $\lambda \in \mathbb{F}$ is then an eigenvalue if and only if $f - \lambda\mathrm{Id}$ is not injective, that is if and only if the kernel does not consist of zero alone.

Definition: If λ is an eigenvalue of f, the subspace

$$E_\lambda := \mathrm{Ker}\,(f - \lambda\mathrm{Id})$$

of V is called the ***eigenspace*** of f for the eigenvalue λ, and its dimension is called the ***geometric multiplicity*** of the eigenvalue.

It is clear that the eigenspaces for distinct eigenvalues only meet in the zero vector, for if $v \neq 0$ and $\lambda \neq \mu$ we cannot have $f(v) = \lambda v = \mu v$. But more is true.

Lemma: If $v_1, \dots, v_r$ are eigenvectors of f for the eigenvalues $\lambda_1, \dots, \lambda_r$, and $\lambda_i \neq \lambda_j$ for $i \neq j$, then $(v_1, \dots, v_r)$ is linearly independent.

PROOF: Start of the induction: the 1-tuple (v_1) is linearly independent, because by definition eigenvectors are nonzero. Inductive assumption: the assertion is correct for $r = k$. Now let $(v_1, \dots, v_{k+1})$ be a $(k+1)$-tuple of eigenvectors for the eigenvalues $\lambda_1, \dots, \lambda_{k+1}$, with $\lambda_i \neq \lambda_j$ for $i \neq j$, and let $\alpha_1 v_1 + \dots + \alpha_{k+1} v_{k+1} = 0$. By applying f respectively multiplication by λ_{k+1} to this relation we obtain the pair of equations

$$\alpha_1 \lambda_1 v_1 + \ldots + \alpha_{k+1} \lambda_{k+1} v_{k+1} = 0$$
$$\alpha_1 \lambda_{k+1} v_1 + \ldots + \alpha_{k+1} \lambda_{k+1} v_{k+1} = 0.$$

By subtraction we have the equation

$$\alpha_1(\lambda_1 - \lambda_{k+1}) v_1 + \dots + \alpha_k(\lambda_k - \lambda_{k+1}) v_k = 0,$$

which no longer contains v_{k+1}, and to which therefore we can apply the inductive assumption stating that $v_1, \dots, v_k$ are linearly independent. Therefore, we have $\alpha_1(\lambda_1 - \lambda_{k+1}) = \dots = \alpha_k(\lambda_k - \lambda_{k+1}) = 0$, and because $\lambda_i \neq \lambda_j$ for $i \neq j$, this implies $\alpha_1 = \dots = \alpha_k = 0$. Hence $\alpha_{k+1} v_{k+1} = 0$, and therefore $\alpha_{k+1} = 0$ also. □

Put somewhat sloppily: eigenvectors for different eigenvalues are linearly independent.

In particular, an endomorphism of an n-dimensional vector space can have at most n distinct eigenvalues, and if it does have this many, it is certainly diagonalizable. This is not, however, a necessary condition for diagonalizability; more precisely we have the following corollary of the lemma.

Corollary: Let $f : V \to V$ be an endomorphism of an n-dimensional vector space over $\mathbb{F}$, let $\lambda_1, \dots, \lambda_r$ be its distinct eigenvalues and $n_1, \dots, n_r$ their geometric multiplicities. Let $(v_1^{(i)}, \dots, v_{n_i}^{(i)})$ be a basis of the eigenspace for λ_i. Then the $(n_1 + \ldots + n_r)$-tuple

$$(v_1^{(1)}, \dots, v_{n_1}^{(1)}, \dots, v_1^{(r)}, \dots, v_{n_r}^{(r)})$$

is linearly independent. In particular, the sum of the geometric multiplicities is *at most* n and f is diagonalizable if and only if this sum *equals* n.

PROOF: If $\sum_{i=1}^{r}\sum_{k=1}^{n_i}\alpha_k^{(i)}v_k^{(i)} = 0$, then by the lemma each of the vectors $\sum_{k=1}^{n_i}\alpha_k^{(i)}v_k^{(i)} \in E_{\lambda_i}$ equals zero, and because $(v_1^{(i)},\ldots,v_{n_i}^{(i)})$ is linearly independent, all the coefficients $\alpha_k^{(i)}$ vanish. Thus by putting bases of all the eigenspaces together, we have indeed obtained a *linearly independent* $(n_1+\cdots+n_r)$-tuple of vectors, which therefore will be a basis in the case that $n_1+\cdots+n_r=n$.

Conversely, if f is assumed to be diagonalizable, then a basis of eigenvectors exists, and if m_i denotes the number of eigenvectors for the eigenvalue λ_i in this basis, then clearly $m_i \le n_i$, hence

$$n = m_1+\cdots+m_r \le n_1+\cdots+n_r \le n.$$

From this it follows that $n_1+\cdots+n_r = n$ and as a byproduct that $m_i = n_i$. □

In broad outline we now see how one has to set about finding a basis of eigenvectors. In the first step one looks for all values of $\lambda \in \mathbb{F}$, for which $f-\lambda\mathrm{Id}$ is not injective — these are the eigenvalues. In the second step one determines a basis of the eigenspace $\mathrm{Ker}(f-\lambda_i\mathrm{Id})$ for each eigenvalue $\lambda_1,\ldots,\lambda_r$. Putting these bases together, if f happens to be *in any way* diagonalizable, will give the desired basis of V consisting of eigenvectors of f.

9.2 The Characteristic Polynomial

Again let V be an n-dimensional vector space over $\mathbb{F}$. We pose the practical problem: for which $\lambda \in \mathbb{F}$ is the endomorphism $f-\lambda\mathrm{Id} : V \to V$ not injective, hence λ an eigenvalue? From the dimension formula for linear maps we know that the rank and kernel dimension add together to give n, so that $f-\lambda\mathrm{Id}$ has a nontrivial kernel if and only if $\mathrm{rk}(f-\lambda\mathrm{Id}) < n$, and this, as we know, will be the case if and only if the *determinant* of $f-\lambda\mathrm{Id}$ vanishes.

Fact 1: If $f : V \to V$ is an endomorphism of a finite-dimensional vector space over $\mathbb{F}$, then $\lambda \in \mathbb{F}$ is an eigenvalue of f if and only if $\det(f-\lambda\mathrm{Id}) = 0$.

In order to work out the determinant, one chooses some basis of V and considers the $n\times n$ matrix A associated to f:

$$\begin{array}{ccc} V & \xrightarrow{\ f\ } & V \\ {\scriptstyle\Phi}\uparrow{\scriptstyle\cong} & & {\scriptstyle\cong}\uparrow{\scriptstyle\Phi} \\ \mathbb{F}^n & \xrightarrow{\ A\ } & \mathbb{F}^n \end{array}$$

(If, as frequently is the case, V is already equal to $\mathbb{F}^n$, this step is of course unnecessary, since we can work with the canonical basis, with respect to which f is already given as a matrix A.) The matrix of $f - \lambda\mathrm{Id}$ is then $A - \lambda E$, where E is the unit matrix. One thus obtains $A - \lambda E$ from A by subtracting λ from each diagonal term. Using the definition of the determinant of an endomorphism (see Section 6.7), one then has the following fact.

Fact 2: If f is decribed by the matrix A with respect to some basis of V, then

$$\det(f - \lambda\,\mathrm{Id}) = \det\begin{pmatrix} a_{11} - \lambda & a_{12} & \dots & a_{1n} \\ a_{21} & a_{22} - \lambda & \dots & a_{2n} \\ \vdots & \vdots & & \vdots \\ a_{n1} & a_{n2} & \dots & a_{nn} - \lambda \end{pmatrix}.$$

This determinant, which depends on λ and which interests us because of the eigenvalues, is called the *characteristic polynomial* P_f of f.

Lemma and Definition: If $f : V \to V$ is an endomorphism of an n-dimensional vector space over $\mathbb{F}$, then there exist coefficients $a_0, \dots, a_{n-1} \in \mathbb{F}$ with

$$\det(f - \lambda\mathrm{Id}) = (-1)^n\lambda^n + a_{n-1}\lambda^{n-1} + \dots + a_1\lambda + a_0 =: P_f(\lambda)$$

for all $\lambda \in \mathbb{F}$. P_f is called the ***characteristic polynomial*** of f.

PROOF: If A and B are $n \times n$ matrices, then $\det(A - \lambda B)$ has the form

$$\det(A - \lambda B) = c_n\lambda^n + \dots + c_1\lambda + c_0$$

for suitable $c_0, \dots, c_n \in \mathbb{F}$. This follows immediately by induction on n, using the expansion formula for the determinant: the start of the induction ($n = 1$) is trivial, and if we expand $\det(A - \lambda B)$ by the first column, for example, then the ith summand is just

$$(-1)^{i+1}(a_{i1} - \lambda b_{i1})\det(A_{i1} - \lambda B_{i1})$$

(see the expansion formula in Section 6.1), and we can apply the inductive assumption to $\det(A_{i1} - \lambda B_{i1})$.

Hence it only remains to show that in the special case $B := E$, occurring in Fact 2, we may take $c_n = (-1)^n$. But this also follows by induction, again using expansion by the first column: the first summand $(a_{11} - \lambda)\det(A_{11} - \lambda E_{11})$, where A_{11} and E_{11} are obtained from A and E by deleting the first row and column, is inductively of the form $(-1)^n\lambda^n$ + terms of lower degree in λ. The other summands $(-1)^{1+i}a_{i1}\det(A_{i1} - \lambda E_{i1})$, by the remarks above on $A - \lambda B$, also involve only lower powers of λ, and so $\det(A - \lambda E) = \det(f - \lambda\mathrm{Id})$ is of the required form. □

Corollary: The eigenvalues are the zeros of the characteristic polynomial.

For $\mathbb{F} = \mathbb{R}$ or $\mathbb{C}$ and $n = 2$, the zeros of the characteristic polynomial $P_f(\lambda) = \lambda^2 + a_1\lambda + a_0$ are easy to determine: since $\lambda^2 + a_1\lambda + a_0 = (\lambda + \frac{a_1}{2})^2 - \frac{a_1}{4}^2 + a_0$, they equal the two numbers $\lambda_{1,2} = -\frac{1}{2}(a_1 \pm \sqrt{a_1^2 - 4a_0})$, as we may say, although strictly speaking in the case $a_1^2 = 4a_0$ there is only one eigenvalue, and if $\mathbb{F} = \mathbb{R}$ and $a_1^2 < 4a_0$, no eigenvalue at all. In concrete applications n is very often equal to 2.

There is, of course, a lot more to say about polynomials in general, but for the specific aims of the present first-year linear algebra text we need to know only one thing: the so-called *fundamental theorem of algebra.*

Fundamental Theorem of Algebra: Each complex polynomial of degree $n \geq 1$, that is, each map $P : \mathbb{C} \to \mathbb{C}$ of the form

$$P(z) = c_n z^n + \cdots + c_1 z + c_0,$$

with $n \geq 1$, $c_0, \ldots, c_n \in \mathbb{C}$, and $c_n \neq 0$, has at least one zero.

For a proof, see [1].

Corollary: For $n \geq 1$ each endomorphism of an n-dimensional complex vector space has at least one eigenvalue.

In Chapter 10 we will use this result in the proof of the theorem about the *principal axes transformation* for a self-adjoint endomorphism of a Euclidean vector space (and in particular for a symmetric matrix).

9.3 Test

(1) In order to be able to discuss the "eigenvalues" of a linear map $f : V \to W$ at all, f must be

- ☐ epimorphic (surjective)
- ☐ isomorphic (bijective)
- ☐ endomorphic ($V = W$)

(2) The vector $v \neq 0$ is called an eigenvector for the eigenvalue λ if $f(v) = \lambda v$. If instead $f(-v) = \lambda v$, then

- ☐ $-v$ is an eigenvector for the eigenvalue λ.
- ☐ v is an eigenvector for the eigenvalue $-\lambda$.
- ☐ $-v$ is an eigenvector for the eigenvalue $-\lambda$.

(3) If $f : V \to V$ is an endomorphism and λ is an eigenvalue of f, then by the eigenspace E_λ of f corresponding to the eigenvalue λ, one understands

☐ the set of all eigenvectors for the eigenvalue λ
☐ the set consisting of all eigenvectors for the eigenvalue λ, together with the zero vector
☐ Ker $(\lambda \mathrm{Id})$

(4) Which of the following three vectors is an eigenvector of
$f = \begin{pmatrix} 2 & 1 \\ 0 & 1 \end{pmatrix} : \mathbb{R}^2 \to \mathbb{R}^2$?

☐ $\begin{pmatrix} 2 \\ 1 \end{pmatrix}$ ☐ $\begin{pmatrix} 1 \\ 1 \end{pmatrix}$ ☐ $\begin{pmatrix} 2 \\ -2 \end{pmatrix}$

(5) Let $f : V \to V$ be an endomorphism of a finite-dimensional vector space, and let $\lambda, \ldots, \lambda_r$ be the distinct eigenvalues of f. Then

☐ $\dim E_{\lambda_1} + \cdots + \dim E_{\lambda_r} = \lambda_1 + \cdots + \lambda_r$.
☐ $\dim E_{\lambda_1} + \cdots + \dim E_{\lambda_r} \leq n$.
☐ $\dim E_{\lambda_1} + \cdots + \dim E_{\lambda_r} > n$.

(6) Let $f : V \xrightarrow{\cong} V$ be an automorphism of V and λ an eigenvalue of f. Then

☐ λ is also an eigenvalue of f^{-1}.
☐ $-\lambda$ is an eigenvalue of f^{-1}.
☐ $\frac{1}{\lambda}$ is an eigenvalue of f^{-1}.

(7) An endomorphism f of an n-dimensional vector space is diagonalizable if and only if

☐ f has n distinct eigenvalues.
☐ f has only one eigenvalue whose geometric multiplicity equals n.
☐ n equals the sum of the geometric multiplicities of the eigenvalues.

(8) The concepts of eigenvalue, eigenvector, eigenspace, geometric multiplicity, and diagonalizability have been defined for endomorphisms of (sometimes finite-dimensional) vector spaces V. Which further "general assumption" on V have we tacitly made here?

☐ V is always a real vector space.
☐ V is always a Euclidean vector space.
☐ no extra assumption; V is just a vector space over $\mathbb{F}$.

(9) The characteristic polynomial of $f = \begin{pmatrix} 1 & 3 \\ -2 & 0 \end{pmatrix} : \mathbb{C}^2 \to \mathbb{C}^2$ is given by

☐ $P_f(\lambda) = \lambda^2 + \lambda + 6$.
☐ $P_f(\lambda) = \lambda^2 - \lambda + 6$.
☐ $P_f(\lambda) = -\lambda + 7$.

(10) If $f, g : V \to V$ are endomorphisms and there exists some $\varphi \in GL(V)$ with $f = \varphi g \varphi^{-1}$, then f and g have

- ☐ the same eigenvalues
- ☐ the same eigenvectors
- ☐ the same eigenspaces

9.4 Polynomials

A section for mathematicians

If $\mathbb{F}$ is an arbitrary field and one considers polynomials in one "indeterminate" λ as expressions of the form $P(\lambda) := c_n\lambda^n + \cdots + c_1\lambda + c_0$, where $n \geq 0$ and the c_i belong to $\mathbb{F}$, then one must distinguish between the ***polynomial*** $P(\lambda)$ and the ***polynomial map*** $P : \mathbb{F} \to \mathbb{F}$ defined by it. This is not just pedantry; it can genuinely happen that polynomials with distinct coefficients $c_0, \ldots, c_n$ and $\widetilde{c}_0, \ldots, \widetilde{c}_m$ give the same polynomial map. Here is an example: if $\mathbb{F} = \mathbb{F}_2 = \{0, 1\}$ is the field of two elements introduced in Section 2.3, then the two polynomials $P(\lambda) := \lambda$ and $\widetilde{P}(\lambda) := \lambda^2$ define the same polynomial map $\mathbb{F}_2 \to \mathbb{F}_2$, since $0 \cdot 0 = 0$ and $1 \cdot 1 = 1$ both hold. One can produce many other examples in this way, and analogously for other finite fields. But for fields with *infinitely* many elements we have the following lemma.

Lemma for Equating Coefficients: If $\mathbb{F} = \mathbb{R}$ or $\mathbb{C}$ (more generally, a field with infinitely many elements), and $P : \mathbb{F} \to \mathbb{F}$ has the form

$$P(\lambda) = c_n\lambda^n + \cdots + c_1\lambda + c_0$$

with coefficients $c_0, \ldots, c_n \in \mathbb{F}$, then these coefficients $c_0, \ldots, c_n$ are uniquely determined by the map P.

Definition: If in addition $c_n \neq 0$, then P is called a polynomial of ***degree*** n.

For $\mathbb{F} = \mathbb{R}$ or $\mathbb{C}$ you may well already have come across the lemma in a calculus course. But it really belongs to the theory of systems of linear equations. If one chooses $n+1$ distinct elements $\lambda_1, \ldots, \lambda_{n+1} \in \mathbb{F}$, the $n+1$ equations

$$c_n\lambda_i^n + \cdots + c_1\lambda_i + c_0 = P(\lambda_i)$$

form a *uniquely* solvable linear system for the unknowns $c_n, \ldots, c_0$. This is because the coefficient matrix of the system has nonzero determinant:

$$\det\begin{pmatrix} \lambda_1^n & \cdots & \lambda_1 & 1 \\ \vdots & & \vdots & \vdots \\ \lambda_{n+1}^n & \cdots & \lambda_{n+1} & 1 \end{pmatrix} = \prod_{i<j}(\lambda_j - \lambda_i)$$

(VANDERMONDE determinant — sneaky inductive proof). Therefore, the coefficients $c_0, \dots, c_n$ of a polynomial of degree at most n are uniquely determined by its values at $n+1$ distinct points. In particular, a polynomial of degree n can have no more than n zeros.

In what follows, the field $\mathbb{F}$ is $\mathbb{R}$ or $\mathbb{C}$, or more generally a field with infinitely many elements.

Lemma: If $P(\lambda)$ is a polynomial of degree $n, n \geq 1$, and $\lambda_0 \in \mathbb{F}$ is one of its zeros, then

$$P(\lambda) = (\lambda - \lambda_0)Q(\lambda)$$

for some well-determined polynomial Q of degree $n-1$.

PROOF: $P(\lambda + \lambda_0)$ is clearly a polynomial of degree n in λ with a zero at 0, hence of the form

$$P(\lambda + \lambda_0) = a_n\lambda^n + \cdots + a_1\lambda = \lambda \cdot (a_n\lambda^{n-1} + \cdots + a_1).$$

Substituting $\lambda - \lambda_0$ for λ, we have

$$P(\lambda) = (\lambda - \lambda_0)(a_n(\lambda - \lambda_0)^{n-1} + \cdots + a_1) =: (\lambda - \lambda_0)Q(\lambda). \qquad \square$$

In practice, a better way to determine the coefficients $b_{n-1}, \dots, b_0$ of Q is directly to compare the coefficients of $P(\lambda)$ and $(\lambda - \lambda_0)Q(\lambda)$. From

$$c_n\lambda^n + \cdots + c_0 = (\lambda - \lambda_0)(b_{n-1}\lambda^{n-1} + \cdots + b_0)$$

we first read off $b_{n-1} = c_n$, and then with the recursion formula

$$b_{k-1} = c_k + \lambda_0 b_k$$

work our way downward to get $b_{n-1}, b_{n-2}, \dots, b_0$ (called "division of P by the linear factor $(\lambda - \lambda_0)$").

If Q also has a zero, we can again split off a linear factor, and so we continue until the process stops. From the fundamental theorem of algebra it therefore follows that over the *complex* numbers one can completely decompose a polynomial into linear factors. More precisely, we have the following corollary.

Corollary and Definition: Each complex polynomial P splits into linear factors, that is, if $P(\lambda) = c_n\lambda^n + \cdots + c_0$ with $c_0 \neq 0$, and $\lambda_1, \dots, \lambda_r \in \mathbb{C}$ are the (pairwise distinct) zeros of P, we have

$$P(\lambda) = c_n \prod_{i=1}^{r} (\lambda - \lambda_i)^{m_i}$$

with well-determined exponents $m_i \geq 1$, called the ***multiplicities*** of the zeros.

In particular, if V is an n-dimensional complex vector space, $f : V \to V$ an endomorphism, and $\lambda_1, \ldots, \lambda_r$ its distinct eigenvalues, we have

$$P_f(\lambda) := \det(f - \lambda \mathrm{Id}) = (-1)^n \prod_{i=1}^{r} (\lambda - \lambda_i)^{m_i} .$$

The m_i are called the ***algebraic multiplicities*** of the eigenvalues λ_i, in contrast to the *geometric* multiplicities $n_i := \dim(\mathrm{Ker}\,(f - \lambda_i \mathrm{Id}))$, the dimensions of the eigenspaces.

We always have $n_i \leq m_i$, since if one extends a basis of an eigenspace, say for the eigenvalue λ_1, to a basis of V, then with respect to this basis f will be represented by a matrix of the form

$$A = \left[\begin{array}{c|c} \begin{matrix} \lambda_1 & & \\ & \ddots & \\ & & \lambda_1 \end{matrix} & * \\ \hline 0 & * \end{array}\right]$$
$$\underbrace{\qquad\quad}_{n_1}$$

and therefore the factor $(\lambda - \lambda_1)$ appears in the linear factorization of $P_f(\lambda) = \det(A - \lambda E)$ at least n_1 times.

The geometric multiplicity may be genuinely smaller than the algebraic; the shear matrix

$$\begin{pmatrix} 1 & 1 \\ 0 & 1 \end{pmatrix}$$

provides an example. The (unique) eigenvalue $\lambda = 1$ has geometric multiplicity 1 and algebraic multiplicity 2.

The sum of the algebraic multiplicities is clearly the degree n of P_f. Since we already know (see Section 9.1) that an endomorphism is diagonalizable if and only if the sum of its *geometric* multiplicities equals n, it follows from $n_i \leq m_i$ that the following condition holds.

Remark: An endomorphism of a finite-dimensional complex vector space is diagonalizable if and only if the geometric multiplicities of its eigenvalues agree with the algebraic multiplicities.

9.5 Exercises

Exercises for mathematicians

9.1: Determine the eigenvalues and associated eigenspaces for the following 2×2 matrices over both the fields $\mathbb{F} = \mathbb{R}$ and $\mathbb{F} = \mathbb{C}$:

$$\begin{pmatrix} 0 & 0 \\ 0 & 0 \end{pmatrix}, \begin{pmatrix} 0 & 1 \\ 1 & 0 \end{pmatrix}, \begin{pmatrix} 0 & 1 \\ 0 & 0 \end{pmatrix}, \begin{pmatrix} 0 & 1 \\ 4 & 3 \end{pmatrix}, \begin{pmatrix} 0 & -1 \\ 1 & 0 \end{pmatrix}, \begin{pmatrix} 0 & 1 \\ -5 & 4 \end{pmatrix}.$$

9.2: A subspace U of V is said to be invariant under an endomorphism f if $f(U) \subset U$. Show that the eigenspaces of $f^n := f \circ \cdots \circ f$ are invariant under f.

9.3: Let $\mathbb{R}^{\mathbb{N}}$ denote the vector space of real sequences $(a_n)_{n \geq 1}$. Determine the eigenvalues and eigenspaces of the endomorphism $f : \mathbb{R}^{\mathbb{N}} \to \mathbb{R}^{\mathbb{N}}$ given by

$$(a_n)_{n \geq 1} \longmapsto (a_{n+1})_{n \geq 1}.$$

Two $*$-exercises

9.1*: Since we can both add and compose endomorphisms of V it makes sense to use the polynomial $P(t) = a_0 + a_1 t + \cdots + a_n t^n$, $a_i \in \mathbb{F}$ to define an endomorphism $P(f) = a_0 + a_1 f + \cdots + a_n f^n : V \to V$. Show that if λ is an eigenvalue of f, then $P(\lambda)$ is an eigenvalue of $P(f)$.

9.2*: Let $\pi : \{1, \ldots, n\} \to \{1, \ldots, n\}$ be a bijective map (permutation). Let $f_\pi : \mathbb{R}^n \to \mathbb{R}^n$ be defined by $f_\pi(x_1, \ldots, x_n) := (x_{\pi(1)}, \ldots, x_{\pi(n)})$. Determine the set of eigenvalues of f_π.

Exercises for physicists

9.1P: Exercise 9.1 (for mathematicians)

9.2P: Suppose that the endomorphism $A : V \to V$ of the two-dimensional vector space V has only one eigenvalue λ, and let E_λ denote the associated eigenspace. Show that $Aw - \lambda w \in E_\lambda$ for all $w \in V$.

9.3P: Let V be the real vector space of twice differentiable functions $f : \mathbb{R} \to \mathbb{R}$. Determine all the eigenvalues of the second derivative

$$\frac{d^2}{dx^2} : V \to V.$$

CHAPTER 10

The Principal Axes Transformation

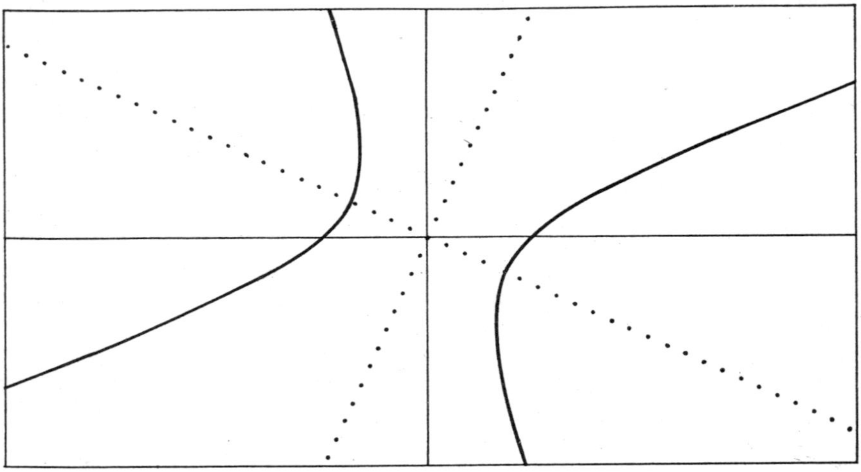

10.1 Self-Adjoint Endomorphisms

The name ***principal axes transformation*** comes from the theory of conic sections. For the plane hyperbola illustrated above, a principal axes transformation would, for example, be an orthogonal map or transformation $P : \mathbb{R}^2 \to \mathbb{R}^2$ that rotates the coordinate axes to the pair of dotted lines marking the two "principal" axes of the hyperbola. However, we will not be concerned with this geometric aspect, but rather with the mathematically equivalent and very important problem of finding an orthonormal basis of eigenvectors for a self-adjoint operator on a finite-dimensional Euclidean vector space.

Definition: Let $(V, \langle\ ,\ \rangle)$ be a Euclidean vector space. An operator or endomorphism $f : V \to V$ is said to be ***self-adjoint*** if $\langle f(v), w\rangle = \langle v, f(w)\rangle$ for all $v, w \in V$.

Two immediate consequences of the self-adjointness condition show that the chances of constructing an orthonormal basis of eigenvectors are good.

Fact 1: Any two eigenvectors v and w of a self-adjoint operator f corresponding to distinct eigenvalues $\lambda \neq \mu$ are orthogonal to each other, since $\langle f(v), w\rangle = \langle v, f(w)\rangle$ implies $\langle \lambda v, w\rangle = \langle v, \mu w\rangle$, thus $(\lambda - \mu)\langle v, w\rangle = 0$.

Fact 2: If v is an eigenvector of the self-adjoint operator $f : V \to V$, then the subspace $v^\perp := \{w \in V \mid w \perp v\}$ is invariant under f, that is, $f(v^\perp) \subset v^\perp$, since $\langle f(w), v\rangle = \langle w, f(v)\rangle = \langle w, \lambda v\rangle$.

Does it not follow at once that we can inductively construct an orthonormal basis of eigenvectors for the n-dimensional vector space V? If v is some eigenvector of the self-adjoint operator $f : V \to V$ and $\dim V = n$, then by inductive assumption there exists an orthonormal basis $(v_1, \ldots, v_{n-1})$ of eigenvectors for the obviously self-adjoint operator

$$f|v^\perp : v^\perp \longrightarrow v^\perp,$$

and need we not only to take $v_n := v/\|v\|$ in order to obtain the desired orthonormal basis $(v_1, \ldots, v_n)$, as illustrated in Fig. 56?

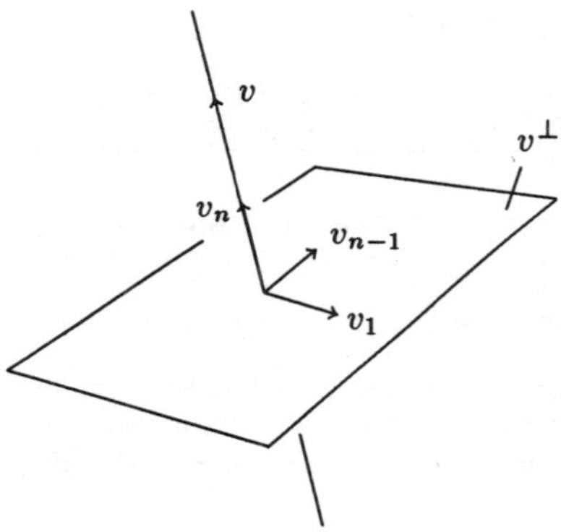

Fig. 56. Finding an orthonormal basis by induction

We *could* argue in this way, if we could only be sure that there always exists "some" eigenvector v in the first place! This is not quite so trivial as the two previous facts, but it does happen to be true and will be proved in Section 10.2.

10.2 Symmetric Matrices

We know that with respect to some basis $(v_1, \ldots, v_n)$ of a vector space V over $\mathbb{F}$, each endomorphism $f : V \to V$ can be described in terms of a matrix $A : \mathbb{F}^n \to \mathbb{F}^n$, related to f by the commutative diagram

$$\begin{array}{ccc} V & \xrightarrow{f} & V \\ \Phi \uparrow \cong & & \Phi \uparrow \cong \\ \mathbb{F}^n & \xrightarrow{A} & \mathbb{F}^n \end{array}$$

in which $\Phi := \Phi_{(v_1,\ldots,v_n)}$, the previously defined basis isomorphism, maps the canonical unit vectors e_j onto $\Phi(e_j) = v_j$. Watching e_j on its two possible routes in the diagram from the lower left to the upper right:

$$\begin{array}{rcccc} \text{first route:} & e_j & \mapsto & v_j & \mapsto & f(v_j) \\ \text{second route:} & e_j & \mapsto & Ae_j & \mapsto & \sum_{i=1}^n a_{ij} v_i, \end{array}$$

(recall that $Ae_j = \sum_{i=1}^n a_{ij}e_i$ is the jth column of A), we conclude that

$$f(v_j) = \sum_{i=1}^n a_{ij}v_i .$$

This quick formula, which is often used to *define* the matrix associated to f, can be applied in *any* n-dimensional vector space. Let us now, more specifically, consider *Euclidean* vector spaces.

Remark: If $(V, \langle\ ,\ \rangle)$ is a Euclidean vector space and $(v_1, \ldots, v_n)$ is an orthonormal basis of V, the matrix A of an endomorphism $f : V \to V$ is given by $a_{ij} = \langle v_i, f(v_j)\rangle$.

PROOF: Expanding $f(v_j)$ in terms of the orthonormal basis, we obtain $f(v_j) = \sum_{i=1}^n \langle v_i, f(v_j)\rangle v_i$, from which the claim follows. □

Corollary: If $(v_1, \ldots, v_n)$ is an orthonormal basis for the Euclidean vector space V, an operator $f : V \to V$ is self-adjoint if and only if its matrix A with respect to $(v_1, \ldots, v_n)$ is symmetric, i.e. $a_{ij} = a_{ji}$.

PROOF: Because $a_{ij} = \langle v_i, f(v_j)\rangle$, the symmetry of A just means that the self-adjointness condition holds for the *basis vectors*, and hence it certainly is necessary. Because of the bilinearity of the inner product it is also sufficient. □

If we want to see this written down, we note that for arbitrary vectors we get $\langle f(v), w\rangle = \langle f(\sum_{i=1}^n x_i v_i), \sum_{j=1}^n y_j v_j\rangle = \sum_{i,j=1}^n x_i y_j \langle f(v_i), v_j\rangle = \sum_{i,j=1}^n x_i y_j \langle v_i, f(v_j)\rangle = \langle v, f(w)\rangle$.

Symmetric matrices and self-adjoint operators on finite-dimensional Euclidean spaces are therefore closely related; in the special case of $V := \mathbb{R}^n$ with the standard inner product they are indeed the same, as one reads off from $\langle Ax, y\rangle = \sum_{ij} a_{ij} x_i y_j$. Thus general statements about self-adjoint operators in finite-dimensional Euclidean spaces always include statements about symmetric matrices and vice versa. We will now use this in the proof of the main technical lemma for the principal axes transformation.

Lemma: Each self-adjoint endomorphism of an n-dimensional Euclidean vector space V with $n > 0$ has an eigenvector.

PROOF: It is enough to prove the theorem for symmetric real $n \times n$ matrices

$$A : \mathbb{R}^n \longrightarrow \mathbb{R}^n$$

since if $A := \Phi \circ f \circ \Phi^{-1}$ is the matrix of a self-adjoint operator $f : V \to V$ with respect to an orthonormal basis, A is symmetric, and if $x \in \mathbb{R}^n$ is an eigenvector of A for the eigenvalue $\lambda \in \mathbb{R}$, then $v := \Phi(x)$ is an eigenvector of f for the same λ.

The eigenvalues are the zeros of the characteristic polynomial. How can we show that there exists $\lambda \in \mathbb{R}$ with $P_A(\lambda) = 0$?

Real polynomials do not need to have zeros in $\mathbb{R}$, but who at this point would not appeal to the one existence theorem for polynomial zeros that we have, namely the fundamental theorem of algebra? So at least there exists a *complex* number

$$\lambda = \gamma + i\omega \in \mathbb{C}$$

with $P_A(\lambda) = 0$. This complex number is therefore an eigenvalue of the endomorphism

$$A : \mathbb{C}^n \longrightarrow \mathbb{C}^n$$

of the *complex* vector space $\mathbb{C}^n$ given by the same matrix A. This means that there exists some nonzero vector

$$z = \begin{pmatrix} z_1 \\ \vdots \\ z_n \end{pmatrix} = \begin{pmatrix} x_1 + iy_1 \\ \vdots \\ x_n + iy_n \end{pmatrix} \in \mathbb{C}^n$$

with $Az = \lambda z$, i.e., $A(x + iy) = (\gamma + i\omega)(x + iy)$, or separating real and imaginary parts:

$$\begin{aligned} Ax &= \gamma x - \omega y \quad \text{and} \\ Ay &= \gamma y + \omega x. \end{aligned}$$

Admittedly, at this moment we do not know whether considering this complex vector will help us, or whether by invoking the complex numbers we have been led astray from our real problem. But until now we have not used the symmetry of the matrix A at all, and before throwing away the two vectors $x, y \in \mathbb{R}^n$ as useless, let us at least write down what the symmetry condition

$$\langle Ax, y \rangle = \langle x, Ay \rangle$$

may have to tell us, namely

$$\langle \gamma x - \omega y, y \rangle = \langle x, \gamma y + \omega x \rangle,$$

thus

$$\gamma \langle x, y \rangle - \omega \langle y, y \rangle = \gamma \langle x, y \rangle + \omega \langle x, x \rangle$$

or

$$\omega \cdot (\|x\|^2 + \|y\|^2) = 0.$$

But since $x + iy = z \neq 0$, it follows from this that $\omega = 0$ and hence $\lambda = \gamma \in \mathbb{R}$. Great! So the characteristic polynomial *has* a real zero after all. □

10.3 The Principal Axes Transformation for Self-Adjoint Endomorphisms

We have already seen indicated in Section 10.1 how the existence of an orthonormal basis of eigenvectors would follow from the existence of eigenvectors for self-adjoint operators. Having proved this existence lemma we can now harvest the following theorem.

Theorem: If $(V, \langle\ ,\ \rangle)$ is a finite-dimensional Euclidean vector space and $f : V \to V$ is a self-adjoint endomorphism, there exists an orthonormal basis of eigenvectors of f.

PROOF: Induction on $n = \dim V$. For $n = 0$ the theorem is trivial (empty basis). Inductive step: let $n \geq 1$. By the lemma there exists an eigenvector v, and by the inductive assumption an orthonormal basis $(v_1, \ldots, v_{n-1})$ of eigenvectors for

$$f|v^{\perp} : v^{\perp} \longrightarrow v^{\perp}.$$

Put $v_n := v/\|v\|$. Then $(v_1, \ldots, v_n)$ is the required basis. □

Corollary (Principal axes transformation for self-adjoint operators): Given a self-adjoint endomorphism $f : V \to V$ of an n-dimensional Euclidean vector space, it is always possible to find an orthogonal transformation

$$P : \mathbb{R}^n \xrightarrow{\cong} V$$

("principal axes transformation"), which reduces f to a diagonal matrix $D := P^{-1} \circ f \circ P$ of the form

$$D = \begin{pmatrix} \lambda_1 & & & & & \\ & \ddots & & & & \\ & & \lambda_1 & & & \\ & & & \ddots & & \\ & & & & \lambda_r & \\ & & & & & \ddots \\ & & & & & & \lambda_r \end{pmatrix}$$

Here $\lambda_1, \ldots, \lambda_r$ are the distinct eigenvalues of f, the number of each appearing on the diagonal being equal to the geometric multiplicity.

In order to obtain such a P we simply take an orthonormal basis of eigenvectors ordered in such a way that eigenvectors for the same eigenvalue are adjacent. The basis isomorphism $\Phi_{(v_1,\ldots,v_n)} =: P$ then has the required property.

In particular, for $V := \mathbb{R}^n$ with the usual inner product we have the following corollary.

Corollary (Principal axes transformation for symmetric real matrices): If A is a symmetric real $n \times n$ matrix, there is an orthogonal transformation $P \in O(n)$, such that $D := P^{-1}AP$ is a diagonal matrix with the eigenvalues of A as diagonal entries, each appearing with its geometric multiplicity.

Anticipating an important generalization, which you may meet in a later course on Functional Analysis, we have the following reformulation.

Corollary (Spectral decomposition of self-adjoint operators): If $f : V \to V$ is a self-adjoint endomorphism of a finite-dimensional Euclidean vector space, $\lambda_1, \dots, \lambda_r$ its distinct eigenvalues, and $P_k : V \to V$ the orthogonal projection onto the eigenspace E_{λ_k}, then

$$f = \sum_{k=1}^{r} \lambda_k P_k .$$

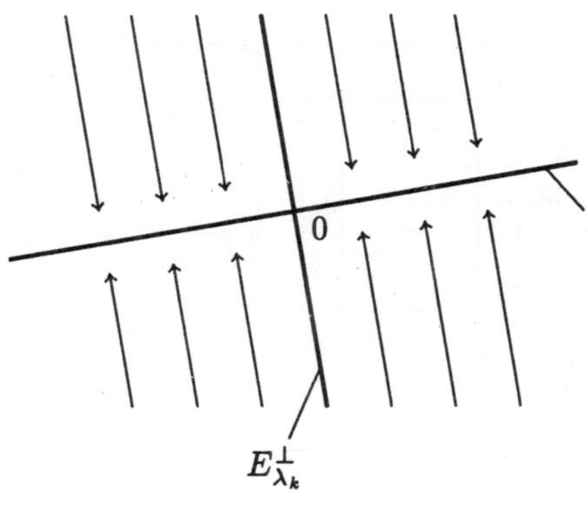

Fig. 57. Orthogonal projection P_k onto the kth eigenspace

PROOF: It suffices to show that both sides have the same effect on eigenvectors of f, since there exists a *basis* of eigenvectors. Suppose, therefore, that v is an eigenvector for the eigenvalue λ_j. Then $f(v) = \lambda_j v$, and

$$P_k(v) = \begin{cases} v & \text{for} \quad k = j \\ 0 & \text{for} \quad k \neq j, \end{cases}$$

because eigenvectors for distinct eigenvalues are orthogonal to each other. Therefore, $\sum_{k=1}^{r} \lambda_k P_k(v) = \lambda_j v = f(v)$. □

If one has to apply the principal axes transformation in practice, the self-adjoint endomorphism is usually already in the form of a symmetric matrix A.

Otherwise, use some arbitrary orthonormal basis to represent f by a symmetric matrix A. The "recipe" for carrying out the principal axes transformation then reads as follows.

> **Recipe:** Let A be a real symmetric $n \times n$ matrix. In order to find a principal axes transformation $P \in O(n)$ for A carry out the following four steps:
>
> FIRST STEP: Form the characteristic polynomial $P_A(\lambda) := \det(A - \lambda E)$ and determine all its zeros, that is, the eigenvalues $\lambda_1, \ldots, \lambda_r$ of A.
>
> SECOND STEP: Determine a basis $(w_1^{(k)}, \ldots, w_{n_k}^{(k)})$ of the eigenspace E_{λ_k} for each $k = 1, \ldots, r$ by applying Gaussian elimination (Section 7.5) to solve the system of equations $(A - \lambda_k E)x = 0$.
>
> THIRD STEP: Orthonormalize $(w_1^{(k)}, \ldots, w_{n_k}^{(k)})$ by the Gram-Schmidt process (Section 8.2), obtaining an orthonormal basis $(v_1^k, \ldots, v_{n_k}^{(k)})$.
>
> FOURTH STEP: Putting these bases together gives an orthonormal basis
>
> $$(v_1, \ldots, v_n) := (v_1^{(1)}, \ldots, v_{n_1}^{(1)}, \ldots, v_1^{(r)}, \ldots, v_{n_r}^{(r)})$$
>
> of V, made up of eigenvectors of A, and we get the desired
>
> $$P := \left(v_1 \,\middle|\, v_2 \,\middle|\, \cdots \,\middle|\, v_n \right)$$
>
> as the matrix with columns $v_1, \ldots, v_n$.

10.4 Test

(1) An endomorphism f of a Euclidean vector space is said to be self-adjoint if, for all $v, w \in V$, we have

- ☐ $\langle f(v), f(w) \rangle = \langle v, w \rangle$.
- ☐ $\langle v, f(w) \rangle = \langle f(v), w \rangle$.
- ☐ $\langle f(v), w \rangle = \langle w, f(v) \rangle$.

(2) If $\lambda_1, \ldots, \lambda_r$ are eigenvalues of a self-adjoint endomorphism, $\lambda_i \neq \lambda_j$ for $i \neq j$, and v_i is an eigenvector for λ_i, $i = 1, \ldots, r$, then for $i \neq j$

☐ $\lambda_i \perp \lambda_j$ ☐ $v_i \perp v_j$ ☐ $E_{\lambda_i} \perp E_{\lambda_j}$

(3) Let V be a finite-dimensional Euclidean vector space. The assertion that for each invariant subspace $U \subset V$ also $U^{\perp}$ is invariant under f holds

☐ for each self-adjoint endomorphism of V
☐ for each orthogonal endomorphism of V
☐ for each endomorphism of V

(4) Which of the following matrices is symmetric?

☐ $\begin{pmatrix} 0 & 0 & 1 & 2 \\ 0 & 0 & 3 & 4 \\ 1 & 2 & 0 & 0 \\ 3 & 4 & 0 & 0 \end{pmatrix}$ ☐ $\begin{pmatrix} 0 & 0 & 1 & 2 \\ 0 & 0 & 3 & 4 \\ 1 & 3 & 0 & 0 \\ 2 & 4 & 0 & 0 \end{pmatrix}$ ☐ $\begin{pmatrix} 1 & 2 & 0 & 0 \\ 3 & 4 & 0 & 0 \\ 0 & 0 & 4 & 2 \\ 0 & 0 & 3 & 1 \end{pmatrix}$

(5) Let A be a real $n\times n$ matrix and $z \in \mathbb{C}^n$ a complex eigenvector, $z = x+iy$ with $x, y \in \mathbb{R}^n$, for the real eigenvalue λ. Suppose that $y \neq 0$. Then

☐ $y \in \mathbb{R}^n$ is an eigenvector of A for the eigenvalue λ.
☐ $y \in \mathbb{R}^n$ is an eigenvector of A for the eigenvalue $i\lambda$.
☐ if $x \neq 0$, then $y \in \mathbb{R}^n$ *cannot* be an eigenvector of A.

(6) For a real symmetric matrix A to carry out the principal axes transformation means finding

☐ a symmetric matrix P so that $P^{-1}AP$ is diagonal
☐ an orthogonal matrix $P \in O(n)$ so that $P^{-1}AP$ is diagonal
☐ an invertible matrix $P \in GL(n, \mathbb{R})$ so that $P^{-1}AP$ is diagonal

(7) Let V be an n-dimensional Euclidean vector space and $U \subset V$ be a k-dimensional subspace. When is the orthogonal projection $P : V \to U$ self-adjoint?

☐ always ☐ only for $0 < k \leq n$ ☐ only for $0 \leq k < n$

(8) Does there exist an inner product on $\mathbb{R}^n$ for which the shear is self-adjoint?

☐ No, because $\begin{pmatrix} 1 & 1 \\ 0 & 1 \end{pmatrix}$ is not diagonalizable.
☐ Yes, let $\langle x, y\rangle := x_1y_1 + x_1y_2 + x_2y_2$.
☐ Yes, because the standard inner product already has this property.

(9) Let $f : V \to V$ be a self-adjoint operator and let $(v_1, \ldots, v_n)$ be a basis of eigenvectors with $\|v_i\| = 1$ for $i = 1, \ldots, n$. Is $(v_1, \ldots, v_n)$ then already an orthonormal basis?

☐ Yes, by definition of an orthonormal basis.
☐ Yes, because the eigenvectors of a self-adjoint operator are orthogonal to each other.
☐ No, because the eigenspaces do not need to be one-dimensional.

(10) If a symmetric real $n \times n$ matrix A has only one eigenvalue λ, then

☐ A is already diagonal.

☐ $a_{ij} = \lambda$ for all $i, j = 1, \ldots, n$.

☐ $n = 1$.

10.5 Exercises

Exercises for mathematicians

10.1: Carry out the principal axes transformation for the symmetric matrix

$$A = \begin{pmatrix} 2 & 1 & 1 \\ 1 & 2 & -1 \\ 1 & -1 & 2 \end{pmatrix};$$

that is, find an orthogonal matrix $P \in O(3)$ so that P^tAP is diagonal.

10.2: Let V be a finite-dimensional real vector space. Show that an endomorphism $f : V \to V$ is diagonalizable if and only if there exists an inner product $\langle\, , \rangle$ on V for which f is self-adjoint.

10.3: Let V be a Euclidean vector space and $U \subset V$ a finite-dimensional subspace. Show that the orthogonal projection $P_U : V \to U$ is self-adjoint, and determine its eigenvalues and eigenspaces.

The $*$-exercise

10*: Let V be a finite-dimensional Euclidean vector space. Show that two self-adjoint endomorphisms $f, g : V \to V$ can be diagonalized by the same orthogonal transformation $P : \mathbb{R}^n \cong V$ if and only if they *commute*, i.e., satisfy $f \circ g = g \circ f$.

Exercises for physicists

10.1P: Exercise 10.1 (for mathematicians)

10.2P: Carry out the principal axes transformation for the symmetric matrix

$$A = \begin{pmatrix} \cos\varphi & \sin\varphi \\ \sin\varphi & -\cos\varphi \end{pmatrix}.$$

10.3P: Determine the dimension of the subspace $\mathrm{Sym}(n, \mathbb{R})$ of $M(n \times n, \mathbb{R})$.

CHAPTER 11

Classification of Matrices

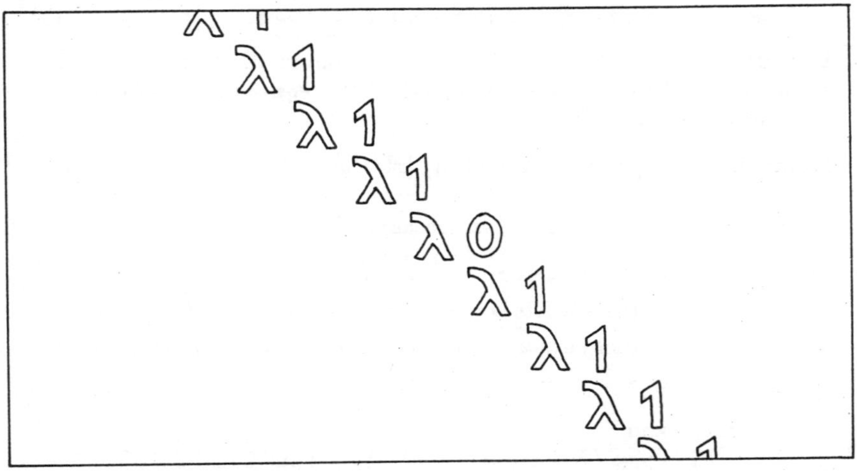

11.1 What Is Meant by "Classification"?

In order to have an overview of a large and perhaps complicated collection of mathematical objects, it is often necessary to ignore certain properties, insofar as these are irrelevant to the matter concerned, and to concentrate on identifying the essentially distinct objects. To some extent it is arbitrary which properties are regarded as "essential"; indeed this depends on the kind of overview for which one is looking. And what does it mean to "ignore" certain properties? And how does one express "essentially the same" and "essentially distinct" in a mathematically sensible way? This last point is the subject of our first definition. We start from the assumption that the objects to be classified form a set M, for example, a set of matrices or a set of subsets of $\mathbb{R}$.

Definition: Let M be a set. By an ***equivalence relation*** on M one understands a relation (formally a subset $R \subset M \times M$, but instead of $(x, y) \in R$ one writes $x \sim y$ and speaks of the "equivalence relation $\sim$"), which satisfies the following three axioms:

(1) Reflexivity: $x \sim x$ for all $x \in M$

(2) Symmetry: $x \sim y \Leftrightarrow y \sim x$

(3) Transitivity: $x \sim y$ and $y \sim x \Rightarrow x \sim z$

Example: Let V be a vector space over $\mathbb{F}$ and $U \subset V$ be a subspace. We define $\sim_U$ on V by $x \sim_U y :\Leftrightarrow x - y \in U$. Then $\sim_U$ is an equivalence relation.

PROOF:

(1) Reflexivity: $x \sim_U x$, since $x - x = 0 \in U$.
(2) Symmetry: $x \sim_U y \Leftrightarrow x - y \in U \Leftrightarrow y - x \in U \Leftrightarrow y \sim_U x$.
(3) Transitivity: $x \sim_U y$ and $y \sim_U z \Rightarrow x - y \in U$ and $y - z \in U$ $\Rightarrow (x - y) + (y - z) \in U \Rightarrow x - z \in U \Rightarrow x \sim z$. □

Definition: Let $\sim$ be an equivalence relation on M. For $x \in M$ the subset $[x] := \{y \mid x \sim y\} \subset M$ is called the ***equivalence class*** of x with respect to $\sim$.

Remark: (i) $\bigcup_{x \in M}[x] = M$ and (ii) $[x] \cap [y] \neq \varnothing \Leftrightarrow x \sim y \Leftrightarrow [x] = [y]$.

PROOF: (i) is trivial since $x \in [x]$ because of axiom 1. For (ii):

$[x] \cap [y] \neq \varnothing \Rightarrow$ there exists some $z \in [x] \cap [y]$
$\Rightarrow$ there exists some z with $x \sim z$ and $y \sim z$
$\Rightarrow$ there exists some z with $x \sim z$ and $z \sim y$ by axiom 2
$\Rightarrow x \sim y$ by axiom 3.

On the other hand:

$x \sim y \Rightarrow (x \sim a \Longleftrightarrow y \sim a)$ by axioms 2 and 3
$\Rightarrow [x] = [y]$.

Finally,

$[x] = [y] \Rightarrow x \in [x] \cap [y]$ by axiom 1
$\Rightarrow [x] \cap [y] \neq \varnothing$. □

For obvious reasons one calls a set of subsets of M, such that each two are disjoint and whose union is M, a *decomposition* of M. The set $\{[x] \mid x \in M\}$ of equivalence classes is an example of such a decomposition.

Example: If U is a subspace of V and $\sim_U$ is defined as above by $x \sim_U y :\Leftrightarrow x - y \in U$, the equivalence classes are called the "cosets" of U, that is, $[x] = x + U = \{x + u \mid u \in U\}$.

Intuitively one may think of such a decomposition as putting the elements of the set into pigeonholes, one pigeonhole for each equivalence class, thus tidying up the originally messy set. This is what classification is about! In order to be more specific, we need names for the set of pigeonholes and for the process of putting the elements into them.

Definition: By the ***quotient*** of M by $\sim$, one understands the set $M/{\sim} := \{[x] \mid x \in M\}$ of equivalence classes, and the map $\pi : M \to M/{\sim}$ mapping x to $[x]$ is called the ***canonical projection*** of the equivalence relation $\sim$.

The choice of an equivalence relation on M is the choice of the point of view from which one wants to obtain a "classification," our strategy for the tidying up of M: equivalent elements go into the same pigeonhole.

But with the choice of the equivalence relation, the classification is not yet done; on the contrary, the work is just starting. We want to get an overview on how many pigeonholes there are, and what is in them and how to decide in which pigeonhole a given element belongs.

What this exactly amounts to is not so easy to explain; "classification" is no rigorously defined term, but it has something of the indeterminacy of the spoken word "overview." But let me try anyway.

Explanation: Let $\sim$ be an equivalence relation on a set M. ***Classifying*** the set M according to $\sim$ or, as one says, ***up to*** $\sim$, amounts to "understanding," "seeing through," "looking over," or "coming to grips with" $M/\sim$, and where possible with $\pi : M \to M/\sim$ also. Two common variants that realize this somewhat vague concept are

(A) classification by means of characteristic data, and
(B) classification by representatives

as follows:

(A) Classification by characteristic data: essentially this consists of finding a "well-known" set D (for example, the set $\mathbb{Z}$ of integers or the like) together with a surjective map $c : M \to D$ with the property $x \sim y \Longleftrightarrow c(x) = c(y)$, which means that the map

$$\begin{aligned} M/\sim \ &\longrightarrow D \\ [x] &\longmapsto c(x) \end{aligned}$$

is well defined and bijective. In this case one says that $c(x)$ is a ***characterizing datum*** for x with respect to $\sim$.

One then "understands" $M/\sim$ and π in the sense that we have a commutative diagram

$$\begin{array}{ccc} M & & \\ {\scriptstyle\pi}\downarrow & \searrow^{c} & \\ M/\sim & \xrightarrow[\cong]{} & D \end{array}$$

in which one understands c and D.

For a simple illustrating example, let M be the set of all finite subsets of $\mathbb{R}^2$. For $X, Y \in M$ we define $X \sim Y :\Longleftrightarrow$ there exists a bijective map $f : X \to Y$. Then one obtains a classification of M according to $\sim$ by characteristic data, if one takes $D := \mathbb{N}$ and $c(X) := \#X$, the number of elements contained in X.

(B) CLASSIFICATION BY REPRESENTATIVES: essentially this consists of finding some "easily understood" subset $M_0 \subset M$ so that

$$\pi|M_0 : M_0 \xrightarrow{\cong} M/\sim$$

is bijective. This means that for each $x \in M$ there exists a unique ***representative*** $x_0 \in M_0$ with $x \sim x_0$. So M_0 contains exactly one sample from each equivalence class. If we can explicitly describe how to *find* this representative for any given $x \in M$, so much the better.

Again, for a simple example, let M be the set of pairs (x, U) consisting of an element $x \in \mathbb{R}^2$ and a one-dimensional subspace U of $\mathbb{R}^2$. In M we introduce an equivalence relation by defining $(x, U) \sim (y, V)$ to mean that there exists an automorphism φ of $\mathbb{R}^2$ with $\varphi(x) = y$ and $\varphi(U) = V$. If $M_0 \subset M$ is the set consisting of the three sample elements $((0,0), \mathbb{R} \times 0)$, $((1,0), \mathbb{R} \times 0)$, and $((0,1), \mathbb{R} \times 0)$, then $\pi|M_0 : M_0 \to M/\sim$ is bijective, and we also know how to find the representative for any given $(x, U) \in M$, since

$$\begin{aligned}(x,U) &\sim ((0,0), \mathbb{R} \times 0) \iff x = 0\\ (x,U) &\sim ((1,0), \mathbb{R} \times 0) \iff x \neq 0, x \in U\\ (x,U) &\sim ((0,1), \mathbb{R} \times 0) \iff x \notin U.\end{aligned}$$

In this way we have arrived at a classification by representatives. Figures 58a, b, and c illustrate the three elements of M_0.

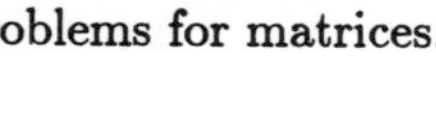

Fig. 58a. First representative

Fig. 58b. Second representative

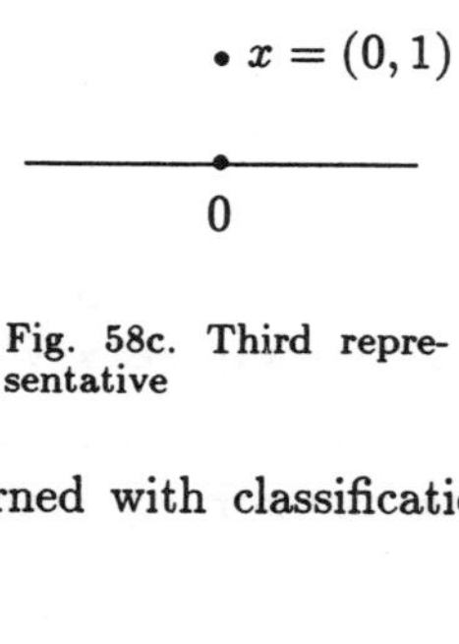

Fig. 58c. Third representative

The following four sections, 11.2–11.5, are concerned with classification problems for matrices.

11.2 The Rank Theorem

Until now we have spoken in a very general way about equivalence. For matrices the word is also used in a more special sense.

Definition: Two $m \times n$ matrices $A, B \in M(m \times n, \mathbb{F})$ are said to be ***equivalent***, written $A \sim B$, if there exist invertible matrices $P \in M(n \times n, \mathbb{F})$ and $Q \in M(m \times m, \mathbb{F})$ with $B = Q^{-1}AP$.

The condition says that the diagram

$$\begin{array}{ccc} \mathbb{F}^n & \xrightarrow{\ A\ } & \mathbb{F}^m \\ {\scriptstyle P}\big\uparrow{\scriptstyle\cong} & & {\scriptstyle\cong}\big\uparrow{\scriptstyle Q} \\ \mathbb{F}^n & \xrightarrow{\ B\ } & \mathbb{F}^m \end{array}$$

is commutative. It is clear that this sort of equivalence defines an equivalence relation on $M(m \times n, \mathbb{F})$: reflexivity, symmetry, and transitivity are easily checked. This is the simplest and coarsest equivalence relation for matrices that has any interest. The associated classification problem is solved by the *rank theorem.*

Rank Theorem: Two $m \times n$ matrices A and B are equivalent in the sense above if and only if they have the same rank.

PROOF: It is clear that equivalent matrices must have the same rank, since then $\operatorname{Im} B$ is mapped by the isomorphism Q onto $\operatorname{Im} A$. Suppose, conversely, that we are given that $\operatorname{rk} A = \operatorname{rk} B$. We construct P and Q in the following manner. First choose a basis $v_1, \ldots, v_{n-r}$ of $\operatorname{Ker} B$ and extend it to a basis $(v_1, \ldots, v_n)$ of all $\mathbb{F}^n$. Then

$$(w_1, \ldots, w_r) := (Bv_{n-r+1}, \ldots, Bv_n)$$

is a basis of $\operatorname{Im} B$, which we can in turn extend to a basis $(w_1, \ldots, w_m)$ of $\mathbb{F}^m$. We do the same thing for A, obtaining the bases $(v'_1, \ldots, v'_n)$ and $(w'_1, \ldots, w'_m)$, respectively, of $\mathbb{F}^n$ and $\mathbb{F}^m$. Now let P and Q be the isomorphisms taking the unprimed into the primed bases. Then $QB = AP$ holds for the basis vectors $v_1, \ldots, v_n$, and hence for all $v \in \mathbb{F}^n$. □

The rank is therefore a characteristic datum for the classification of $m \times n$ matrices up to equivalence, and since all ranks between zero and the maximum possible rank $r_{\max} := \min(m, n)$ do occur, the rank defines a bijection $M(m \times n, \mathbb{F})/\!\sim\ \xrightarrow{\cong}\ \{0, \ldots, r_{\max}\}$.

At the same time we can give a classification by representatives or, as one says, *normal forms.* For example, we can choose the $m \times n$ matrices of the form

$$r\left\{\begin{array}{c|c} \begin{matrix}1 & & \\ & \ddots & \\ & & 1\end{matrix} & 0 \\ \hline 0 & 0 \end{array}\right. =: E_r^{m\times n}$$

with $0 \le r \le r_{\max} = \min(m, n)$ as normal forms, and then each $m \times n$ matrix is equivalent to precisely one of these normal forms, namely to the one with the same rank. To "bring" or to "reduce" A to its normal form means in this context just to find invertible matrices P and Q with $Q^{-1}AP = E_r^{m\times n}$. The proof of the rank theorem shows how this can be done.

11.3 The Jordan Normal Form

If we are interested in $n \times n$ matrices as *endomorphisms* of $\mathbb{F}^n$, then we will not be happy with the coarse classification of the previous section, because it takes no account of the finer properties of endomorphisms, such as the eigenvalues and the characteristic polynomial. The appropriate equivalence relation here is the *similarity* of matrices.

> **Definition:** Two $n \times n$ matrices A, B are said to be ***similar*** if there exists an invertible $n \times n$ matrix P making the diagram
>
> $$\begin{array}{ccc} \mathbb{F}^n & \xrightarrow{\;A\;} & \mathbb{F}^n \\ {\scriptstyle P}\uparrow{\scriptstyle \cong} & & {\scriptstyle \cong}\uparrow{\scriptstyle P} \\ \mathbb{F}^n & \xrightarrow{\;B\;} & \mathbb{F}^n \end{array}$$
>
> commutative, i.e., $B = P^{-1}AP$.

Similarity is again clearly an equivalence relation. Similar matrices are certainly "equivalent" in the sense of the previous section, but the converse is false because, for example, similar matrices must have the same characteristic polynomial.

The classification of $n \times n$ matrices up to similarity is not as simple as the rank theorem, and in this first-year text I will merely state rather than prove the result, and then only for the case $\mathbb{F} := \mathbb{C}$, that is, only for complex $n \times n$ matrices. But even if you will meet the proof only later or, if mathematics is not your major, maybe never, at least the theorem will give you a feel for the nature of complex endomorphisms.

The individual building blocks for normal forms are of the following type.

> **Definition:** Let λ be a complex number and $m \geq 1$. The $m \times m$ matrix
>
> $$J_m(\lambda) := \begin{pmatrix} \lambda & 1 & & \\ & \ddots & \ddots & \\ & & \ddots & 1 \\ & & & \lambda \end{pmatrix}$$
>
> is called the ***Jordan block*** of degree m for the eigenvalue λ.

As an endomorphism of $\mathbb{C}^m$ the Jordan block $J_m(\lambda)$ clearly has only the eigenvalue λ, and the dimension of the eigenspace is the smallest that an eigenspace can have, namely 1. For $m \geq 2$ such a Jordan block therefore is not diagonalizable; indeed one can say that it is as nondiagonalizable as a complex $m \times m$ matrix can be.

Jordan Normal Form Theorem: If A is a complex $n \times n$ matrix, and if $\lambda_1, \ldots, \lambda_r \in \mathbb{C}$ are its distinct eigenvalues, then for each $k = 1, \ldots, r$ there exist uniquely determined positive natural numbers n_k and

$$m_1^{(k)} \le m_2^{(k)} \le \cdots \le m_{n_k}^{(k)}$$

with the property that there exists an invertible complex $n \times n$ matrix P such that $P^{-1}AP$ is the "block matrix" obtained by adjunction of the Jordan blocks

$$J_{m_1^{(1)}}(\lambda_1), \ldots, J_{m_{n_1}^{(1)}}(\lambda_1), \ldots\ldots, J_{m_1^{(r)}}(\lambda_r), \ldots, J_{m_{n_r}^{(r)}}(\lambda_r)$$

along the diagonal.

Thus the kth eigenvalue λ_k contributes a smaller block matrix, let us call it B_k, built up from n_k single Jordan blocks:

$B_k =$

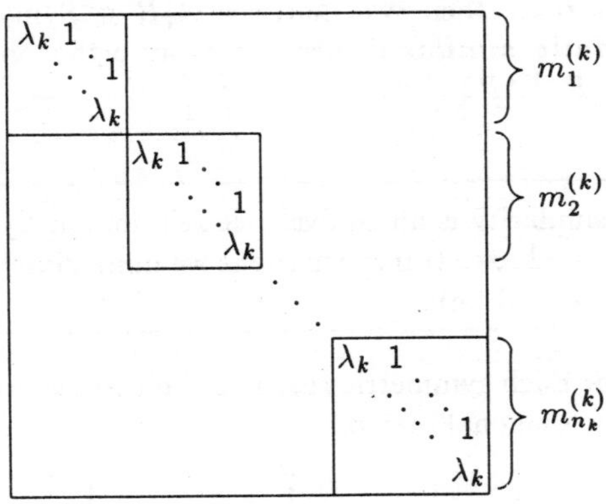

The whole *Jordan normal form* of A is then

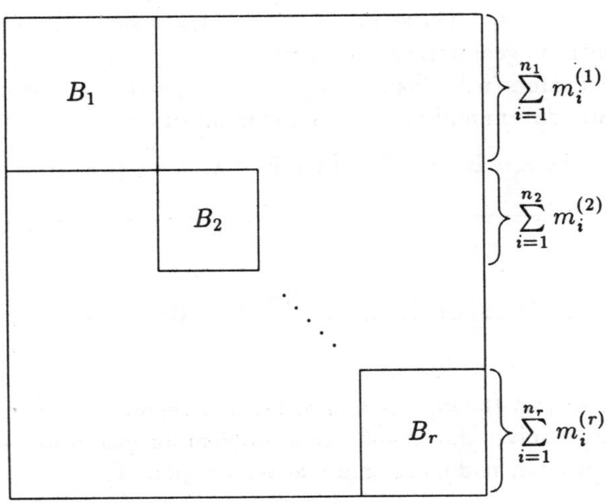

Apart from the fact that there is no prescribed order for the eigenvalues of a complex matrix, the theorem gives us a similarity classification for complex $n \times n$ matrices by representatives in normal form, and the set of eigenvalues together with the information that associates with each eigenvalue the ordered sequence of numbers specifying its Jordan blocks provides characteristic data. Only if each Jordan block has dimension 1 is A diagonalizable.

11.4 More on the Principal Axes Transformation

The theorem on the principal axes transformation for self-adjoint operators in finite-dimensional Euclidean vector spaces solves a further classification problem for matrices, namely that of symmetric real $n \times n$ matrices up to *orthogonal similarity.*

Definition: If $\mathrm{Sym}(n, \mathbb{R})$ denotes the vector space of symmetric real $n \times n$ matrices, then two matrices $A, B \in \mathrm{Sym}(n, \mathbb{R})$ are said to be ***orthogonally similar*** if there exists an orthogonal matrix $P \in O(n)$ with $B = P^{-1}AP$.

Orthogonal similarity is an equivalence relation on $\mathrm{Sym}(n, \mathbb{R})$. From the theorem on principal axes transformations we immediately obtain a classification by means of normal forms.

Theorem: Each symmetric real $n \times n$ matrix is orthogonally similar to exactly one diagonal matrix,

$$A = \begin{pmatrix} \lambda_1 & & \\ & \ddots & \\ & & \lambda_n \end{pmatrix}$$

with $\lambda_1 \leq \cdots \leq \lambda_n$. These $\lambda_1, \ldots, \lambda_n$ are the eigenvalues of A, each occuring with its geometric multiplicity.

The eigenvalues with their geometric multiplicities therefore form a characteristic datum and give a classifying bijection:

$$\mathrm{Sym}(n, \mathbb{R})/\sim \ \xrightarrow{\cong} \ \{\lambda \in \mathbb{R}^n \mid \lambda_1 \leq \cdots \leq \lambda_n\}.$$

11.5 The Sylvester Inertia Theorem

The Sylvester inertia theorem is a classification theorem for symmetric real $n \times n$ matrices; it solves the classification problem for *quadratic forms* on $\mathbb{R}^n$, and therefore on each n-dimensional real vector space V.

Definition: If V is a real vector space and $b : V \times V \to \mathbb{R}$ is a symmetric bilinear form, then

$$\begin{aligned} q : V &\longrightarrow \mathbb{R} \\ v &\longmapsto b(v, v) \end{aligned}$$

is called the ***quadratic form*** on V ***associated*** to b. Note that b can be recovered from q, since the bilinearity of b implies

$$b(v + w, v + w) = b(v, v) + b(w, v) + b(v, w) + b(w, w),$$

and since b is symmetric,

$$b(v, w) = \frac{1}{2}(q(v + w) - q(v) - q(w)).$$

Hence one can also call b the bilinear form ***associated*** to q.

What have quadratic forms to do with matrices? If $(v_1, \ldots, v_n)$ is a basis of V and we write $v = x_1 v_1 + \cdots + x_n v_n$, then again because of the bilinearity and symmetry of b we have

$$q(v) = b(v, v) = b(\sum_i x_i v_i, \sum_j x_j v_j) = \sum_{i,j=1}^{n} b(v_i, v_j) x_i x_j ,$$

and we therefore make the following definitions.

Definition: The symmetric matrix A given by $a_{ij} := b(v_i, v_j)$ is called the ***matrix of the quadratic form*** $q : V \to \mathbb{R}$ with respect to the basis $(v_1, \ldots, v_n)$.

Notation and Remark: For $A \in \mathrm{Sym}(n, \mathbb{R})$ let $Q_A : \mathbb{R}^n \to \mathbb{R}$ denote the quadratic form given by

$$Q_A(x) := \sum_{i,j=1}^{n} a_{ij} x_i x_j .$$

So if A is the matrix of $q : V \to \mathbb{R}$ with respect to $(v_1, \ldots, v_n)$, and Φ is the basis isomorphism $\mathbb{R}^n \cong V$ for this basis, then

$$\begin{array}{ccc} V & \xrightarrow{\;q\;} & \mathbb{R} \\ \Phi \uparrow \cong & \nearrow Q_A & \\ \mathbb{R}^n & & \end{array}$$

is commutative.

Why does one need quadratic forms? Among the many uses of quadratic forms in mathematics I want to single out one for your attention, which you will soon come across in an analysis course. From differential calculus in *one* variable you know what information is carried by the second derivative $f''(x_0)$ concerning the behavior of f near a "critical point" ($f'(x_0) = 0$). If $f''(x_0) > 0$ ($f''(x_0) < 0$), then the function has a local minimum (maximum) at x_0, and the question only remains open if $f''(x_0) = 0$, in which case one needs additional information about the higher derivatives. In differential calculus in *several* variables the situation is similar, but the Taylor polynomial of degree two at a critical point is no longer simply $f(x_0) + (1/2)f''(x_0)\xi^2$, but is in n variables:

$$f(x_0) + \frac{1}{2!}\sum_{i,j=1}^{n} \frac{\partial^2 f}{\partial x_i \partial x_j}(x_0)\xi_i\xi_j,$$

thus given by the constant term $f(x_0)$ and a *quadratic form* with the matrix

$$a_{ij} := \frac{1}{2}\frac{\partial^2 f}{\partial x_i \partial_j}(x_0).$$

In one variable it is easy to see from the sign of $f''(x_0)$ how $\xi \mapsto (1/2)f''(x_0)\xi^2$ behaves. But for the symmetric matrix A, if this does not by chance have a particularly simple form and instead stands there full of numbers, it is at first hard to see what is going on. For this one needs a little quadratic form theory, to which we now turn again.

Definition: Let q be a quadratic form on the n-dimensional real vector space V. A basis $(v_1, \ldots, v_n)$ of V for which the matrix A of q has the form

$$\begin{matrix} r\left\{\begin{matrix} \\ \\ \end{matrix}\right. \\ s\left\{\begin{matrix} \\ \\ \end{matrix}\right. \\ \\ \end{matrix}\begin{pmatrix} +1 & & & & & & \\ & \ddots & & & & & \\ & & +1 & & & & \\ & & & -1 & & & \\ & & & & \ddots & & \\ & & & & & -1 & \\ & & & & & & 0 \\ & & & & & & & \ddots \\ & & & & & & & & 0 \end{pmatrix}$$

($0 \le r, s$ and $r + s \le n$) shall be called a ***Sylvester basis*** for q.

With respect to such a basis the quadratic form takes the simple enough form:

$$q(x_1v_1 + \cdots + x_nv_n) = x_1^2 + \cdots + x_r^2 - x_{r+1}^2 - \cdots - x_{r+s}^2.$$

Sylvester Inertia Theorem: For a quadratic form q on an n-dimensional real vector space V there always exists a Sylvester basis. The numbers r and s of positive and negative entries in the associated diagonal matrix are independent of the choice of the Sylvester basis.

Definition: We say that $r + s$ is the ***rank*** and $r - s$ is the ***signature*** of the quadratic form.

PROOF: (a) Existence of the Sylvester basis: we find such a basis by induction on n, rather as in the principal axes transformation, only this time it is much easier. The theorem is trivial for $q = 0$; let us suppose

$q \neq 0$. Then there must be a vector $v \in V$ with $q(v) = \pm 1$, and this is all we need for the inductive step. If b is the symmetric bilinear form of q, then

$$U := \{w \in V \mid b(v,w) = 0\}$$

is an $(n-1)$-dimensional subspace of V (this follows from the dimension formula for the map $b(v,\cdot) : V \to \mathbb{R}$, $w \mapsto b(v,w)$). By the inductive assumption $q|U$ has a Sylvester basis $(u_1, \ldots, u_{n-1})$, and we only need to add v in the right place in order to obtain a Sylvester basis for all of V.

(b) r and s are well defined: the quantity r can be defined independently of bases as the maximum dimension of a subspace of V on which q is positive definite. In order to see this, take some Sylvester basis and consider the subspaces V_+ and $V_{-,0}$ spanned by the first r and last $n-r$ vectors respectively. Then $q|V_+$ is positive definite, but each higher dimensional subspace U must by the dimension formula meet $V_{-,0}$ non-trivially. Therefore $q|U$ cannot be positive definite. Analogously, s is the maximum dimension of a subspace on which q is negative definite. □

Now consider quadratic forms on $\mathbb{R}^n$. What does it mean for two matrices $A, B \in \mathrm{Sym}(n,\mathbb{R})$ to have quadratic forms that only differ by an isomorphism P of $\mathbb{R}^n$, in the sense that the diagram

$$\begin{array}{ccc} \mathbb{R}^n & \xrightarrow{Q_A} & \\ P \uparrow \cong & & \mathbb{R} \\ \mathbb{R}^n & \xrightarrow{Q_B} & \end{array}$$

commutes, that is $Q_B = Q_A \circ P$? To see this, it is convenient to interpret each n-tuple $x \in \mathbb{R}^n$ as an $n \times 1$ matrix or column, and thus the corresponding transposed $1 \times n$ matrix x^t as a row.

Remark: If one writes $x \in \mathbb{R}^n$ as a column, then the 1×1 matrix $Q_A(x)$ equals the product

$$\begin{aligned} Q_A(x) &= x^t \cdot A \cdot x \\ &= (x_1, \ldots, x_n) \begin{pmatrix} a_{11} & \cdots & a_{1n} \\ \vdots & & \vdots \\ a_{n1} & \cdots & a_{nn} \end{pmatrix} \begin{pmatrix} x_1 \\ \vdots \\ x_n \end{pmatrix}, \end{aligned}$$

and since the rule $(XY)^t = Y^t X^t$ holds for transposed matrices, $Q_B = Q_A \circ P$ is equivalent to $x^t B x = x^t P^t A P x$ being satisfied for all $x \in \mathbb{R}^n$ or

$$B = P^t A P.$$

This describes the effect of a "coordinate transformation" P on the matrix of a quadratic form.

From $Q_B = Q_A \circ P$ we can see that B is just the matrix of Q_A with respect to the basis formed from the columns of P. The Sylvester inertia theorem therefore implies the following corollary.

Corollary: If symmetric real $n \times n$ matrices A and B are called equivalent if their quadratic forms are related by a coordinate transformation, i.e. if there exists $P \in GL(n,\mathbb{R})$ with $B = P^t AP$, then each $A \in \mathrm{Sym}(n,\mathbb{R})$ is equivalent to a unique normal form

$$\begin{matrix} r\left\{\begin{matrix}\\ \\ \\ \end{matrix}\right. \\ s\left\{\begin{matrix}\\ \\ \\ \end{matrix}\right. \\ \\ \end{matrix} \begin{pmatrix} +1 & & & & & & & \\ & \ddots & & & & & & \\ & & +1 & & & & & \\ & & & -1 & & & & \\ & & & & \ddots & & & \\ & & & & & -1 & & \\ & & & & & & 0 & \\ & & & & & & & \ddots \\ & & & & & & & & 0 \end{pmatrix}$$

with $0 \leq r, s$ and $r + s \leq n$.

In this way we obtain a classification by representatives.

In contrast to the classification of symmetric matrices up to orthogonal similarity in Section 11.4, for the equivalence relation now being studied there are, for each fixed n, only finitely many equivalence classes: the pair (r,s) is a charateristic datum and gives us a bijection

$$\mathrm{Sym}(n,\mathbb{R})/\sim \ \xrightarrow{\cong} \ \{(r,s) \mid 0 \leq r, s \text{ and } r + s \leq n\}.$$

Note that orthogonally similar symmetric matrices A and B have equivalent quadratic forms, since $P \in O(n)$ implies that $P^{-1} = P^t$, so $B = P^{-1}AP$ implies that $B = P^t AP$.

If one has used the principal axes transformation to bring a symmetric matrix into diagonal form, one can of course read the Sylvester characteristic data r and s from it: these are the numbers of positive and negative eigenvalues counted with their multiplicities. But there is a much more convenient way of finding the Sylvester normal form, and it avoids the determination of eigenvalues altogether.

For this we recall the observation made in Section 5.5 that elementary row transformations on the left-hand factor of a matrix product XY reproduce themselves on the product matrix. Analogously, the same is true for column transformations on the right-hand factor.

Moreover, if P_2 results from P_1 by an elementary column transformation, then of course P_2^t results from P_1^t through the corresponding *row* transformation, since transposition interchanges rows and columns. For the product matrix $P_1^t AP_1$, passage to $P_2^t AP_2$ thus means that we carry out the column and corresponding row transformations *simultaneously*. We shall refer to this as an ***elementary symmetric transformation***. We now have all the ingredients for the following recipe.

Recipe for finding the Sylvester normal form: If one uses a finite sequence of elementary symmetric transformations to reduce a symmetric real matrix A to some Sylvester normal form,

$$\begin{matrix} r\left\{ \vphantom{\begin{matrix}1\\1\end{matrix}} \right. \\ s\left\{ \vphantom{\begin{matrix}1\\1\end{matrix}} \right. \\ \vphantom{\begin{matrix}1\\1\end{matrix}} \end{matrix} \begin{pmatrix} +1 & & & & & & \\ & \ddots & & & & & \\ & & +1 & & & & \\ & & & -1 & & & \\ & & & & \ddots & & \\ & & & & & -1 & \\ & & & & & & 0 \\ & & & & & & & \ddots \\ & & & & & & & & 0 \end{pmatrix} =: S,$$

then this is in fact *the* Sylvester normal form of A, because if one applies the corresponding column transformations alone to the unit matrix E, one obtains a matrix $P \in GL(n, \mathbb{R})$ with $P^t AP = S$, so the columns of P form a Sylvester basis for A.

If one is interested only in r and s and not in P, then it suffices to use elementary symmetric transformations to bring A into diagonal shape: r and s are the numbers of positive and negative diagonal elements.

11.6 Test

(1) Which of the following properties of an equivalence relation is (are) not fulfilled for the relation on $\mathbb{R}$ defined by $x \leq y$?

☐ Reflexivity ☐ Symmetry ☐ Transitivity

(2) We define an equivalence relation on $\mathbb{Z}$ by "$n \sim m \Leftrightarrow n - m$ is even." How many elements are there in $\mathbb{Z}/\!\sim$?

☐ 1 ☐ 2 ☐ infinitely many

(3) If two $m \times n$ matrices A and B have the same rank, there exist invertible matrices X and Y so that

☐ $AX = BY$ ☐ $AX = YB$ ☐ $XA = YB$

(4) Are the 2×2 matrices $A = \begin{pmatrix} 1 & 2 \\ 0 & 1 \end{pmatrix}$ and $B = \begin{pmatrix} 2 & 4 \\ 0 & 2 \end{pmatrix}$ similar?

☐ Yes, because $B = 2A$.
☐ Yes, because they have the same rank.
☐ No, because they have distinct eigenvalues.

(5) How does the Jordan normal form of $A = \begin{pmatrix} 2 & 3 & 4 \\ 0 & 2 & 5 \\ 0 & 0 & 2 \end{pmatrix}$ look?

□ $\begin{pmatrix} 2 & & \\ & 2 & \\ & & 2 \end{pmatrix}$ □ $\begin{pmatrix} 2 & 0 & \\ & 2 & 1 \\ & & 2 \end{pmatrix}$ □ $\begin{pmatrix} 2 & 1 & \\ & 2 & 1 \\ & & 2 \end{pmatrix}$

(6) Is the Jordan normal form of a real symmetric $n \times n$ matrix always diagonal?

- □ Yes, because it can be brought into diagonal form by means of the principal axes transformation.
- □ No, because a symmetric matrix can also have fewer than n distinct eigenvalues. The argument in favor of the answer "Yes" is unsound, since $O(n) \neq GL(n, \mathbb{C})$.
- □ The question has no meaning and therefore does not deserve an answer, since the Jordan normal form theorem is stated not for real but for *complex* $n \times n$ matrices.

(7) Starting from a quadratic form $q : V \to \mathbb{R}$ one obtains the associated bilinear symmetric form b from

- □ $b(v, w) = \frac{1}{4}(q(v + w) - q(v - w))$.
- □ $b(v, w) = \frac{1}{2}(q(v) + q(w))$.
- □ $b(v, w) = \frac{1}{2}(q(v) + q(w) - q(v - w))$.

(8) What is the matrix for the quadratic form $q : \mathbb{R}^3 \to \mathbb{R}$ given by $q(x, y, z) = 4x^2 + 6xz - 2yz + 8z^2$?

□ $\begin{pmatrix} 4 & 0 & 6 \\ 0 & 0 & -2 \\ 6 & -2 & 8 \end{pmatrix}$ □ $\begin{pmatrix} 4 & 0 & 3 \\ 0 & 0 & -1 \\ 3 & -1 & 8 \end{pmatrix}$ □ $\begin{pmatrix} 2 & 0 & 3 \\ 0 & 0 & -1 \\ 3 & -1 & 4 \end{pmatrix}$

(9) Is the rank $r + s$ of a quadratic form $q : V \to \mathbb{R}$ equal to the rank of the matrix A representing q with respect to some basis?

- □ Yes, since $r + s$ is the rank of the Sylvester matrix S of q, and we have $S = P^t AP$.
- □ No, $r + s$ is only the maximal rank of a matrix representing q.
- □ No, the rank of A is $r - s$, because s is the number of entries equal to -1.

(10) Let A be a symmetric real 2×2 matrix with $\det A < 0$. How does the Sylvester normal form of A look?

- □ $\begin{pmatrix} -1 & \\ & -1 \end{pmatrix}$
- □ $\begin{pmatrix} +1 & \\ & -1 \end{pmatrix}$
- □ We cannot decide this from $\det A < 0$ alone. We need at least one further piece of information in order to determine the *two* quantities r and s.

11.7 Exercises

Exercises for mathematicians

11.1: The proof of the rank theorem in Section 11.2 is given in terms of the column rank and makes no use of the fact that this coincides with the row rank. Show that the rank theorem implies this agreement as a corollary.

Hint: the equality $(XY)^t = Y^t X^t$ for transposed matrices (see the remark in Section 6.3) follows, without relying on the rank theorem, directly from the definition of transposition. Therefore one may use it here.

11.2: In the theorem in Section 11.3 about the Jordan normal form we are explicitly given a Jordan matrix with eigenvalues $\lambda_1, \dots, \lambda_r$ and Jordan blocks with degrees $m_1^{(k)} \leq \cdots \leq m_{n_k}^{(k)}$ for the eigenvalue λ_k. Determine the geometric and algebraic multiplicities of the eigenvalues of this matrix.

11.3: Use symmetric transformations to determine an invertible 3×3 matrix P so that for

$$A = \begin{pmatrix} 1 & 1 & 1 \\ 1 & 0 & 1 \\ 1 & 1 & 1 \end{pmatrix}$$

$P^t AP$ is in Sylvester normal form.

The $*$-exercise

11*: A property of real $n \times n$ matrices shall be called "open" if all matrices A with this property form an open subset of $M(n \times n, \mathbb{R}) = \mathbb{R}^{n^2}$. Determine the openness or nonopenness of the following properties:

(a) invertibility
(b) symmetry
(c) diagonalizability
(d) $\text{rk}\,(A) \leq k$
(e) $\text{rk}\,(A) \geq k$

Exercises for physicists

11.1P: For the symmetric real 5×5 matrix

$$A = \begin{pmatrix} 1 & 1 & 1 & 1 & 1 \\ 1 & 1 & 1 & 1 & 1 \\ 1 & 1 & 1 & 1 & 1 \\ 1 & 1 & 1 & 1 & 1 \\ 1 & 1 & 1 & 1 & 1 \end{pmatrix},$$

determine the diagonal matrix that would result from applying the principal axes transformation, and find the Sylvester normal form.

Given the particular simplicity of the matrix A, do not calculate but merely *think* (what is the image of A, what does A do to vectors in $\operatorname{Im} A$, etc.).

11.2P: Determine the rank and signature of the quadratic form defined on $\mathbb{R}^3$ by $Q(x,y,z) := x^2 + 8xy + 2y^2 + 10yz + 3z^2 + 12xz$.

11.3P: Exercise 11.3 (for mathematicians).

Chapter 12
Answers to the Tests

The charts indicating the correct answers are followed by comments to the individual questions. These comments should help to close those gaps in knowledge revealed by an incorrect answer.

Chapter 1 Test

1	2	3	4	5	6	7	8	9	10
×						×		×	
				×	×		×		
	×	×	×						×

(1) Read the definition on page 4.

(2) Look at the figures on page 3.

(3) Read the text on page 2 (for the definition of $\{a\}$), read the definition of the Cartesian product on page 4, look at the diagram on page 4, and let b move while keeping a fixed.

(4) "Constant" does not mean that a map "does nothing" (as one could perhaps say of Id_M), but that all $x \in M$ are mapped to a single point. Reread the definition of a constant map on page 7.

(5) Reexamine the definition of projection onto the first factor (page 7) and if necessary, also that of the Cartesian product on page 4.

(6) Reread the definitions of $f(A)$ and $f^{-1}(B)$ on page 7.

(7) Read the definition of gf on page 9. The expression $g(x)$ is not even defined, since g is defined on Y, and the third answer is meaningless.

(8) One can only follow the arrows in two ways going from X to Y. Read the definition of a commutative diagram on page 9.

(9) f^{-1} must map $1/x$ to x, therefore $1/x$ to $1/(1/x)$, therefore t to $1/t$.

(10) Not injective since, for example, $(-1)^2 = (+1)^2$; not surjective since, for example, $-1 \neq x^2$ for all $x \in \mathbb{R}$. Reread the definitions of injective and surjective on page 8.

CHAPTER 2 TEST

1	2	3	4	5	6	7	8	9	10
					×		×		
×			×					×	
	×	×		×		×			×

(1) Reread the definition of $\mathbb{R}^n$ in Chapter 1, page 5.

(2) No vector multiplication is involved in the definition of a vector space.

(3) Multiply out $(x - yi)(a + bi)$ and observe that $i^2 = -1$.

(4) "Scalar multiplication" (not to be confused with the "inner" or "scalar" product defined on a Euclidean vector space; see Section 2.4) does not mean the multiplication of scalars with each other (in which case $\mathbb{F} \times \mathbb{F} \to \mathbb{F}$ would be the correct answer), but of scalars with vectors, thus a map $\mathbb{F} \times V \to V$.

(5) Read the definition of a real vector space on page 17 again. Answer 2 does not make sense, but it is important to realize the distinction between the first and the third answers.

(6) Recall the definition of $X \times Y$ in Chapter 1.

(7) Each subspace of V must contain the zero of V (compare the proof of the remark in Section 2.3).

(8) In all three examples it is true that $U \neq \varnothing$, and in the first two it is also true that $\lambda x \in U$ if $x \in U$ and $\lambda \in \mathbb{R}$, but only the first satisfies $x + y \in U$ for all $x, y \in U$.

(9) In the complex plane $\mathbb{C}$ the imaginary numbers $iy = (0, y)$ form the y-axis, and therefore it is not true that $U = \mathbb{C}$. The third answer would have been correct, if it had been asked whether U were a subspace of the *complex* vector space $\mathbb{C}$.

(10) Each line through the origin is also a subspace, not just the two axes.

Chapter 3 Test

1	2	3	4	5	6	7	8	9	10
×				×				×	
	×		×			×			
		×			×		×		×

(1) Reread the definition of linear independence on page 44.

(2) If necessary reread the definitions of basis (page 45) and dimension (page 46).

(3) From $\lambda_1 v_1 + \lambda_2 v_2 = 0$ it always follows that $\lambda_1 v_1 + \lambda_2 v_2 + 0 \cdot v_3 = 0$, hence $\lambda_1 = \lambda_2 = 0$.

(4) $(e_1, \cdots, e_n)$ in the third answer would also be a basis, but not the canonical basis. In case you chose the first answer, read the definition of $\mathbb{F}^n = \mathbb{F} \times \cdots \times \mathbb{F}$ on page 5 and then that of a basis, page 45.

(5) As we have often argued, $0 \cdot v = 0$ (see, for example, page 24) and $1 \cdot v = v$ (axiom 6 on page 17). Therefore, the second statement says nothing about $v_1, \ldots, v_n$, and the third implies that $v_1 = \cdots = v_n = 0$.

(6) Reread the basis extension theorem on page 46, and if necessary reflect on the statements about the 0-tuple $\varnothing$ on page 43.

(7) Combine the definition of a basis on page 45 with the statement about the 0-tuple on page 43.

(8) We do have $y \in U_2 \Leftrightarrow -y \in U_2$; therefore, $U_1 - U_2$ would again be nothing else but the set of all sums $x + (-y)$ of an element from U_1 with one from U_2.

(9) In order to have a simple example showing that the second and third answers are false, let $V = \mathbb{R}^2$ and let U_1, U_2, and U_3 be the two axes and their bisector.

(10) Here the dimension formula says that $\dim U_1 + \dim U_2 = n + \dim U_1 \cap U_2$ (Theorem 3, page 49), from which it follows that the third answer is correct. The falsity of the first two follows already from the example $V = U_1 = U_2 = \mathbb{R}$.

CHAPTER 4 TEST

1	2	3	4	5	6	7	8	9	10
×		×	×		×				
		×		×			×	×	
	×				×	×			×

(1) If we let $\lambda = \mu = 1$ (or $\mu = 0$), the linearity of f follows from the first assertion (compare the definition on page 62). Conversely, from linearity it follows that $f(\lambda x + \mu y) = f(\lambda x) + f(\mu y) = \lambda f(x) + \mu f(y)$. The other two answers make no sense.

(2) Even someone who knows only what a linear map is, and who has forgotten the definition of $\operatorname{Ker} f$, can guess why the first two answers will be wrong. $f(\lambda x) = \lambda f(x)$ implies $f(0) = 0$ (let $\lambda = 0$), hence according to the first two answers $\operatorname{Ker} f$ would be equal to $\{0\}$ for all linear f. Compare to Fact 3 and Definition on page 63.

(3) Here there are two correct answers. The last does not make sense, since $f(\lambda)$ is not even defined. $f(-x) = -f(x)$ follows from $-v = (-1)v$ (see page 43).

(4) The first assertion implies that f is bijective, hence that f is an isomorphism (compare to Exercise 1.2). The third answer would be correct if instead of writing "each n-tuple" one were to write "each basis" (compare to Remark 3 on page 66). The second answer says nothing about f.

(5) Of course one cannot guess this, one has to *know* it. See the definition on page 67.

(6) Reread pages 69 and 70.

(7) If we write (x, y) as a column, we have

$$\begin{pmatrix}1\\0\end{pmatrix} \mapsto \begin{pmatrix}1\\-1\end{pmatrix} \quad \text{and} \quad \begin{pmatrix}0\\1\end{pmatrix} \mapsto \begin{pmatrix}1\\1\end{pmatrix},$$

giving the matrix $\begin{pmatrix}1 & 1\\-1 & 1\end{pmatrix}$, since the columns of the matrix are the images of the unit vectors.

(8)

$$\begin{array}{ccc} v_i & \overset{f}{\longmapsto} & w_i \\ \cong\uparrow & & \uparrow\cong \\ e_i & \longmapsto & e_i \end{array}$$

Reread the definitions on pages 72 and 73 and the text between them.

(9) The right answer comes out of Fact 3 (page 63). The first answer resembles Fact 5 (page 68), but for this one must presuppose that V and W have the same finite dimension.

(10) The dimension formula for maps gives it, for here $n = 5$ and $\operatorname{rk} f = 3$.

Chapter 5 Test

1	2	3	4	5	6	7	8	9	10
×	×	×					×	×	×
			×	×		×			
					×				

(1) Reread the definition of matrix addition on page 85.

(2) Compare with test question (8) from Chapter 4: $\begin{pmatrix} 1 & 0 & 0 \\ 0 & 1 & 0 \\ 0 & 0 & 1 \end{pmatrix}$ is the matrix of the identity $\mathbb{F}^3 \to \mathbb{F}^3$, and the composition of any map with the identity always gives this map. If one wants to see this in "matrix language," one applies matrix multiplication (see Figure 39 on page 87) to our case. One obtains counterexamples to the other answers by taking B to be equal to the matrix of the identity.

(3) I am always hesitating myself. Perhaps the following helps: the first index is the row index, and correspondingly the m in $M(m \times n, \mathbb{F})$ gives the number of rows.

(4) Multiplication of matrices. Look at the definition on page 87, and read the text on pages 87 and 88. Learn it by heart as soon as possible.

(5) Does everything become clear if I spell out what the three words mean? Associativity: $(AB)C = A(BC)$; commutativity: $AB = BA$; distributivity: $A(B + C) = AB + AC$ and $(A + B)C = AC + BC$ whenever these sums and products are defined. Not yet? Reread the remark on page 88.

(6) For a matrix $A \in M(n \times n, \mathbb{F})$ of rank n, the map $A : \mathbb{F}^n \to \mathbb{F}^n$ is surjective (see the definition on page 90). Now recall that for linear maps between $\mathbb{F}^n$ and $\mathbb{F}^n$ injectivity and surjectivity are the same (Fact 5 on page 68).

(7) $\operatorname{rk} \begin{pmatrix} 1 & 1 \\ 2 & 7 \end{pmatrix} = 2$, but $\operatorname{rk} \begin{pmatrix} 1 & 4 \\ 2 & 8 \end{pmatrix} = 1$; now compare with Remark 1 on page 91.

(8) From $BA = E$ it follows that A is injective and B surjective, because $Ax = Ay \Rightarrow BAx = BAy \Leftrightarrow x = y$, and $y = B(Ay)$. However, B does not need to be injective, nor A surjective, and $m > n$ can genuinely happen. Example: let $m = 3, n = 2, A(x_1, x_2) := (x_1, x_2, x_3)$, and $B(x_1, x_2, x_3) := (x_1, x_2)$.

(9) Recall the concept of linear inependence in Chapter 3. What does the "maximum number of linearly independent columns" mean? (Compare with the definition on page 90.)

(10) Rank = row rank = maximum number of linearly independent rows: always smaller than or equal to the number of rows, therefore $\operatorname{rk} A \le m$. (Compare with the definition, fact and theorem on page 90.)

CHAPTER 6 TEST

1	2	3	4	5	6	7	8	9	10
	×					×	×		×
×		×							
	×		×	×	×			×	

(1) Compare Theorem 1 and the definition on pages 103 and 104.

(2) The determinant is invariant under row transformations of type (3) (compare the definition on page 91), but type (1) alters the sign. For type (2) (multiplication of a row by $\lambda \neq 0$) the determinant is multiplied by λ (compare the lemma on page 104). Therefore the first and third statements are correct.

(3) Compare the lemma on page 112.

(4) Counterexample to the first answer: $A = B = \begin{pmatrix} 1 & 0 \\ 0 & 1 \end{pmatrix}$, because then $\det(A + B) = \det \begin{pmatrix} 2 & 0 \\ 0 & 2 \end{pmatrix} = 4$, but $\det A + \det B = 2$. Remark on the second: $\det \lambda A = \lambda^n \det A$, for multiplying only *one* row by λ already multiplies the determinant by λ. The correctness of the last answer follows from Theorem 4, page 112.

(5) The first answer gives a valid equation, but it refers to expansion by the jth *column*. Reread the two expansion formulae on pages 107 and 110.

(6) Expanding by the first column (see page 107) gives

$$\det A = 1 \cdot \det \begin{pmatrix} 3 & -1 \\ 1 & 1 \end{pmatrix} - 2 \cdot \det \begin{pmatrix} 0 & 1 \\ 1 & 1 \end{pmatrix} = 1 \cdot 4 - 2 \cdot (-1) = 6.$$

(7) Definition on page 109.

(8)

$$\operatorname{rk}\begin{pmatrix} \lambda & \lambda & \lambda \\ \lambda & \lambda & \lambda \\ \lambda & \lambda & \lambda \end{pmatrix} \leq 1 \Longrightarrow \det \begin{pmatrix} \lambda & \lambda & \lambda \\ \lambda & \lambda & \lambda \\ \lambda & \lambda & \lambda \end{pmatrix} = 0.$$

(9) $\cos^2 \varphi + \sin^2 \varphi = 1$.

(10) Counterexample to the first statement: $A = \begin{pmatrix} 2 & 1 \\ 1 & 1 \end{pmatrix} \neq \begin{pmatrix} 1 & 0 \\ 0 & 1 \end{pmatrix} = E$, but $\det A = 2 - 1 = 1$. For the other two answers, see the lemma in Section 6.5.

Chapter 7 Test

1	2	3	4	5	6	7	8	9	10
			×		×				
	×			×			×		
×		×			×	×		×	×

(1) The first two examples are also systems of linear equations, but of a very special kind and not written in the usual way. See page 120.

(2) $A \in M(m \times n, \mathbb{F})$ has m rows and n columns, therefore by the definition on page 69 it carries a map $A : \mathbb{F}^n \to \mathbb{F}^m$. Hence $Ax \in \mathbb{F}^m$ for $x \in \mathbb{F}^n$; anything else doesn't work.

(3) See page 121.

(4) In the augmented matrix

$$(A, b) = \left(\begin{array}{c|c} \quad A \quad & b \end{array}\right)$$

b is then linearly superfluous (see the beginning of the proof that row rank = column rank, page 90), and therefore $\operatorname{rk} A = \operatorname{rk}(A, b)$ (by Remark 1 on page 121). If, moreover, the jth column of A equals b, then clearly $x = (0, \ldots, 1, \ldots, 0)$, with the 1 in the jth position, is a solution of $Ax = b$.

(5) Why should $Ax = b$ be always solvable? For $A = 0 \in M(n \times n, \mathbb{F})$ and $b \neq 0$, for example, assuredly not. For n equations in n unknowns, Remark 1 on page 121 applies as well.

(6) For $A \in M(n \times n, \mathbb{F})$, we have:

$$\dim \operatorname{Ker} A = 0 \Leftrightarrow \operatorname{rk} A = n \Leftrightarrow A : \mathbb{F}^n \to \mathbb{F}^n \text{ is bijective}$$

(compare with the commentary for Question (6) in Test 5). On the other hand, $\dim \operatorname{Ker} A = n$ simply means that $A = 0$.

(7) Counterexample to the first two answers: let $A = \begin{pmatrix} 1 & 1 \\ 1 & 1 \end{pmatrix}$. Then $Ax = b$ is solvable for $b = \begin{pmatrix} 2 \\ 2 \end{pmatrix}$, but not for $b = \begin{pmatrix} 1 \\ 2 \end{pmatrix}$. $Ax = b$ can be *uniquely* solvable only if $\operatorname{Ker} A = \{0\}$; see Remark 2 on page 121.

(8) If $\dim \operatorname{Ker} A = 1$, it is still possible for $x_0 + \operatorname{Ker} A$ to contain two linearly independent elements, as shown in Figure 59.

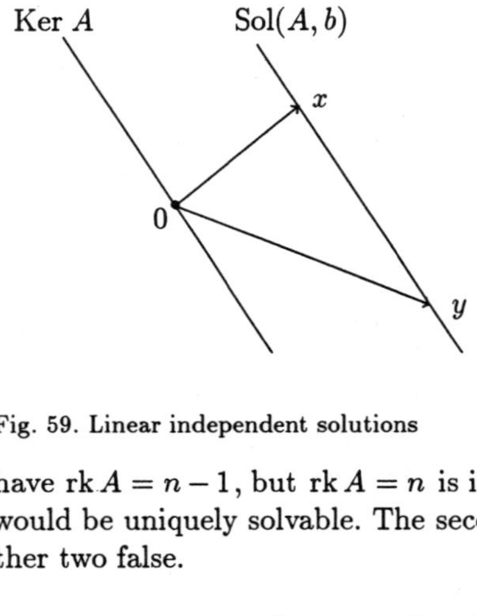

Fig. 59. Linear independent solutions

Therefore we can have $\operatorname{rk} A = n - 1$, but $\operatorname{rk} A = n$ is impossible, since in this case $Ax = b$ would be uniquely solvable. The second answer is thus correct, and the other two false.

(9) For $A = 0$ one cannot carry out the first step, but impossibility to do the first step does not imply that the whole matrix vanishes, only that the first column must be zero.

(10) The first condition is not necessary, the second is neither necessary (example $\begin{pmatrix} 1 & 1 \\ 1 & 0 \end{pmatrix}$) nor sufficient (example $\begin{pmatrix} 1 & 1 & 0 \\ 1 & 1 & 1 \\ 0 & 1 & 1 \end{pmatrix}$).

Chapter 8 Test

1	2	3	4	5	6	7	8	9	10
×							×		
			×	×		×			×
	×	×			×			×	

(1) Compare with the definition on page 136.

(2) The second assertion is certainly a consequence of positive definiteness, but is not equivalent to it. Consider property (iii) on page 136.

(3) Given the answer to Question (1), the second statement is meaningless, and the first is a mistake frequently made by beginners. There are many inner products for $\mathbb{R}^n$ other than the one given by $\langle x, y\rangle = x_1y_1 + \ldots x_ny_n$.

(4) See the third of the four definitions on page 139.

(5) Since $\dim \mathbb{R}^2 = 2$ the third answer must be wrong. For the first we do have $\langle(1,-1),(-1,-1)\rangle = 0$, but $||(1,-1)|| = ||(-1,-1)|| = \sqrt{2} \neq 1$. Compare to the last of the four definitions on page 139.

(6) Apart from physicists, who have already worked this out as an exercise in Chapter 4, you will not find it so easy to see the correctness of the third answer. It follows from $\langle x, y\rangle = (1/4)(||x+y||^2 - ||x-y||^2)$. The second answer, although not correct, is not so far off the mark. For such maps we always have $f(x) = \lambda\varphi(x)$ for all $x \in V$ and some $\lambda \in \mathbb{R}$ and a suitable orthogonal map φ.

(7) Orthogonal maps are always injective (see the fact on page 143). Therefore, we must have $\operatorname{Ker} P_U = U^\perp = 0$, if P_U is to be orthogonal. By the corollary on page 143 we then have $U = V$, and $P_V = \operatorname{Id}_V$ is certainly orthogonal.

(8) See the fact on page 144.

(9) The second answer is not as wrong as the first, since $(\mathbb{Z}, +)$ would indeed be a group, but the formulation is not acceptable. $\mathbb{N}$ is not unfit to have *some* group structure $\mathbb{N} \times \mathbb{N} \to \mathbb{N}$, but since "+" here denotes the ordinary addition of numbers, axiom (3) is not satisfied.

(10) A $2k \times 2k$ matrix is not a $k \times k$ matrix, so the first answer must be wrong. The statement $(-1)^{2k} = 1$ in the third is only a decoy, quite irrelevant to the question. $SO(2k) \neq O(2k)$ because, for example,

$$\begin{pmatrix} -1 & & & \\ & 1 & & \\ & & \ddots & \\ & & & 1 \end{pmatrix} \in O(2k) \setminus SO(2k).$$

CHAPTER 9 TEST

1	2	3	4	5	6	7	8	9	10
									×
	×	×		×				×	
×	×		×		×	×	×		

(1) If $f(x) = \lambda x$ is to mean anything, x and $f(x)$ must belong to the same vector space. Look at the definition on page 151.

(2) $f(-x) = \lambda x \Rightarrow f(-x) = (-\lambda)(-x)$ and $f(x) = (-\lambda)x$, so x and $-x$ are eigenvectors for $-\lambda$.

(3) $E_\lambda = \operatorname{Ker}(f - \lambda \mathrm{Id})$ contains zero as well as the eigenvectors for λ.

(4) $\begin{pmatrix} 2 & 1 \\ 0 & 1 \end{pmatrix} \begin{pmatrix} 2 \\ -2 \end{pmatrix} = \begin{pmatrix} 2 \\ -2 \end{pmatrix}$, therefore $\begin{pmatrix} 2 \\ -2 \end{pmatrix}$ is an eigenvector for the eigenvalue $\lambda = 1$.

(5) See the corollary on page 153.

(6) $f(x) = \lambda x \Rightarrow x = f^{-1}(\lambda x) = \lambda f^{-1}(x) \Rightarrow f^{-1}(x) = \frac{1}{\lambda}x$. Note that λ *is* in fact nonzero, otherwise f would fail to be injective and could not be an automorphism.

(7) All three conditions imply diagonalizability (see page 153, the corollary on page 153, and, of course, the definition of diagonalizability on page 152), but the first two are not *equivalent* to diagonalizability, as the example

$$f = \begin{pmatrix} 1 & 0 & 0 \\ 0 & 1 & 0 \\ 0 & 0 & 0 \end{pmatrix} : \mathbb{R}^3 \to \mathbb{R}^3$$

shows.

(8) We have only restricted $\mathbb{F}$ to be $\mathbb{R}$ or $\mathbb{C}$ on page 159, so as to avoid certain technicalities with polynomials over arbitrary fields.

(9) $\det \begin{pmatrix} 1-\lambda & 3 \\ -2 & -\lambda \end{pmatrix} = (1-\lambda)(-\lambda) - 3(-2) = \lambda^2 - \lambda + 6$.

(10) An easy calculation shows that v is an eigenvector of f for the eigenvalue λ, if and only if $\varphi^{-1}(v)$ is an eigenvector of g for the eigenvalue λ.

CHAPTER 10 TEST

1	2	3	4	5	6	7	8	9	10
		×		×		×	×		×
×	×	×	×		×				
	×							×	

(1) Endomorhpisms having the first property are called orthogonal, while the third property imposes no conditions on f (symmetry of the inner product).

(2) The λ_i are real numbers, not elements of V, therefore the first statement is meaningless. For the correctness of the two others, see Fact 1 on page 162.

(3) The same argument as used in Fact 2 on page 162 shows that the first two answers are both correct: from $u \in U$ and $w \in U^{\perp}$ it follows that $\langle f(w), u\rangle = \langle w, f(u)\rangle = 0$ for a self-adjoint f, and $\langle f(w), u\rangle = \langle f^{-1}f(w), f^{-1}(u)\rangle = \langle w, f^{-1}(u)\rangle = 0$ for an orthogonal f. (Note that in the latter case $f: V \to V$ and $f|U: U \to U$ are even isomorphisms.)

(4) For the first matrix $a_{14} \neq a_{41}$, for the third $a_{12} \neq a_{21}$.

(5) Because A and λ are real, it follows from $A(x+iy) = \lambda(x+iy)$ that $Ax = \lambda x$ and $Ay = \lambda y$. The second answer only holds in the special case $\lambda = 0$.

(6) Look at the corollary on page 167.

(7) $\langle P_U v, w\rangle = \langle P_U v, P_U w\rangle$, because $P_U v \in U$, and the vectors w and $P_U w$ have the same U-component. Hence $\langle P_U v, w\rangle = \langle v, P_U w\rangle$, and the first answer is correct; see Chapter 8, page 142.

(8) The first argument is sound, so one doesn't really need to work out the second. The third is quite wrong, since $\begin{pmatrix} 1 & 1 \\ 0 & 1 \end{pmatrix}$ is not symmetric.

(9) The second answer would be correct if it were additionally assumed that the eigenvalues $\lambda_1, \dots, \lambda_n$ are all distinct. But without this assumption the third step in the recipe on page 168 is not superfluous.

(10) By means of the principal axes transformation we certainly have $P^{-1}AP = \lambda E$, and therefore $A = P\lambda EP^{-1} = \lambda E$. The second answer would only be true for $\lambda = 0$.

CHAPTER 11 TEST

1	2	3	4	5	6	7	8	9	10
					×	×		×	
×	×	×					×		×
			×	×		×			

(1) We have $x \leq x$, and from $x \leq y \leq z$ it follows that $x \leq z$, but $y \leq x$ does not follow from $x \leq y$: the symmetry requirement is not satisfied.

(2) There are precisely two equivalence classes: the first consists of the odd numbers, the second of the even numbers.

(3) The second condition says that A and B are equivalent in the sense of the definition on page 174; therefore, by the rank theorem the second answer is correct. From the first condition it would follow that A and B have the same image, from the third that they have the same kernel, neither being deducible from the equality of ranks.

(4) It is true that $B = 2A$ and that $\operatorname{rk} A = \operatorname{rk} B = 2$, but neither implies that $B = P^{-1}AP$. Because $\det(P^{-1}AP - \lambda E) = \det(P^{-1}(A - \lambda E)P) = (\det P)^{-1} \det(A - \lambda E) \det P = \det(A - \lambda E)$, the characteristic polynomials of A and $P^{-1}AP$ are the same. But the characteristic polynomial of A is $(1 - \lambda)^2$ and that of B is $(2 - \lambda)^2$.

(5) The characteristic polynomial of A is $P_A(\lambda) = (2 - \lambda)^3$, so $\lambda = 2$ is the unique eigenvalue. Hence only the three given matrices can come into consideration for the Jordan normal form. But the dimension of the eigenspace $\operatorname{Ker}(A - 2E)$ is one, since

$$\begin{pmatrix} 0 & 3 & 4 \\ & 0 & 5 \\ & & 0 \end{pmatrix}$$

has rank 2. This must also be the case for the Jordan normal form, and so only the third answer is correct.

(6) If some Jordan block in the Jordan normal form of $A \in \operatorname{Sym}(n, \mathbb{R})$ were to have degree ≥ 2, then A would not be diagonalizable; see page 176. Just *because* of $O(n) \subset GL(n, \mathbb{R}) \subset GL(n, \mathbb{C})$, the argument is valid.

We are not going to accept the third answer, which is pretending not to know of $\mathbb{R} \subset \mathbb{C}$.

(7) Note that $q(v + w) = b(v + w, v + w) = q(v) + 2b(v, w) + q(w)$.

(8) If we write x_1, x_2, x_3 instead of x, y, z, we have $q(x) = \sum_{i,j=1}^{3} a_{ij} x_i x_j$, which because of $a_{ij} = a_{ji}$ is the same as

$$\sum_{i=1}^{3} a_{ii} x_i^2 + \sum_{i<j} 2a_{ij} x_i x_j .$$

Therefore the coefficients of the squares are the matrix elements a_{ii}, but those of the mixed terms are $(a_{ij} + a_{ji}) = 2a_{ij}$.

(9) Indeed it is as the first answer says, since S and A both describe $q : V \to \mathbb{R}$ with respect to suitable bases, which means $q \circ \Phi = q_S$ and $q \circ \Psi = q_A$ for the corresponding basis isomorphisms; hence $q_S = q_A \circ \Psi^{-1} \circ \Phi =: q_A \circ P$; see page 179.

(10) If $\begin{pmatrix} \lambda_1 & 0 \\ 0 & \lambda_2 \end{pmatrix}$ is the diagonal form of A after applying the principal axes transformation, $\det A = \lambda_1 \lambda_2$, therefore A has one positive and one negative eigenvalue. From this it follows that $r = s = 1$ (see page 182).

References

[1] Ebbinghaus, H. D. et al., *Numbers*. Springer-Verlag: New York, 1990.

[2] Gericke, H., *Geschichte des Zahlbegriffs*. BI Hochschultaschenbuch 172/172a*: Mannheim, Germany, 1970.

[3] Halmos, P. R., *Finite-dimensional Vector Spaces,* Second Edition. D. van Nostrand: Princeton, New Jersey, 1958.

[4] Kittel, C., Knight, W. D., and Ruderman, M. A., *Mechanics*. Berkeley Physics Course, Vol. I. McGraw-Hill: New York, 1965.

[5] Schwerdtfeger, H., *Introduction to Linear Algebra and the Theory of Matrices*. P. Noordhoff N.V.: Groningen, Netherlands, 1961.

[6] Steinitz, E., *Bedingt konvergente Reihen und konvexe Systeme*. J. Reine Angew. Math., 143(1913), 128-175.

[7] *Neun Bücher Arithmetischer Technik. Ein chinesisches Rechenbuch für den praktischen Gebrauch aus der frühen Hanzeit (202 v. Chr. bis 9 n. Chr.), übersetzt und erläutert von Kurt Vogel.* Friedr. Vieweg & Sohn: Braunschweig, Germany, 1968.

Index

(continued)

Lang: Undergraduate Algebra. Second edition.
Lang: Undergraduate Analysis.
Lax/Burstein/Lax: Calculus with Applications and Computing. Volume 1.
LeCuyer: College Mathematics with APL.
Lidl/Pilz: Applied Abstract Algebra.
Macki-Strauss: Introduction to Optimal Control Theory.
Malitz: Introduction to Mathematical Logic.
Marsden/Weinstein: Calculus I, II, III. Second edition.
Martin: The Foundations of Geometry and the Non-Euclidean Plane.
Martin: Transformation Geometry: An Introduction to Symmetry.
Millman/Parker: Geometry: A Metric Approach with Models. Second edition.
Owen: A First Course in the Mathematical Foundations of Thermodynamics.
Palka: An Introduction to Complex Function Theory.
Pedrick: A First Course in Analysis.
Peressini/Sullivan/Uhl: The Mathematics of Nonlinear Programming.
Prenowitz/Jantosciak: Join Geometries.
Priestley: Calculus: An Historical Approach.
Protter/Morrey: A First Course in Real Analysis. Second edition.
Protter/Morrey: Intermediate Calculus. Second edition.
Ross: Elementary Analysis: The Theory of Calculus.
Samuel: Projective Geometry.
Readings in Mathematics.
Scharlau/Opolka: From Fermat to Minkowski.
Sigler: Algebra.
Silverman/Tate: Rational Points on Elliptic Curves.
Simmonds: A Brief on Tensor Analysis. Second edition.
Singer/Thorpe: Lecture Notes on Elementary Topology and Geometry.
Smith: Linear Algebra. Second edition.
Smith: Primer of Modern Analysis. Second edition.
Stanton/White: Constructive Combinatorics.
Stillwell: Elements of Algebra: Geometry, Numbers, Equations.
Stillwell: Mathematics and Its History.
Strayer: Linear Programming and Its Applications.
Thorpe: Elementary Topics in Differential Geometry.
Troutman: Variational Calculus with Elementary Convexity.
Valenza: Linear Algebra: An Introduction to Abstract Mathematics.